Praxis Core Math Test Prep

The Ultimate Guide to Praxis Core Math + 2 Full-Length Practice Tests

By

Reza Nazari

Copyright © 2023

Effortless Math Education Inc.

All rights reserved. No part of this publication may be reproduced, stored in a retrieval system, or transmitted in any form or by any means, electronic, mechanical, photocopying, recording, scanning, or otherwise, except as permitted under Section 107 or 108 of the 1976 United States Copyright Ac, without permission of the author.

Effortless Math provides unofficial test prep products for a variety of tests and exams. It is not affiliated with or endorsed by any official organizations.

All inquiries should be addressed to:

info@effortlessMath.com

www.EffortlessMath.com

ISBN: 978-1-63719-117-0

Published by: Effortless Math Education Inc.

www.EffortlessMath.com

Visit www.EffortlessMath.com

for Online Math Practice

Welcome to

Praxis Core Math Prep 2023

Thank you for choosing Effortless Math for your Praxis Core Math test preparation and congratulations on making the decision to take the Praxis Core test! It's a remarkable move you are taking, one that shouldn't be diminished in any capacity. That's why you need to use every tool possible to ensure you succeed on the test with the highest possible score, and this extensive study guide is one such tool.

If math has never been a strong subject for you, don't worry! This book will help you prepare for (and even ACE) the Praxis Core test's math section. As test day draws nearer, effective preparation becomes increasingly more important. Thankfully, you have this comprehensive study guide to help you get ready for the test. With this guide, you can feel confident that you will be more than ready for the Praxis Core Math test when the time comes.

First and foremost, it is important to note that this book is a study guide and not a textbook. It is best read from cover to cover. Every lesson of this "self-guided math book" was carefully developed to ensure that you are making the most effective use of your time while preparing for the test. This up-to-date guide reflects the 2023 test guidelines and will put you on the right track to hone your math skills, overcome exam anxiety, and boost your confidence, so that you can have your best to succeed on the Praxis Core Math test.

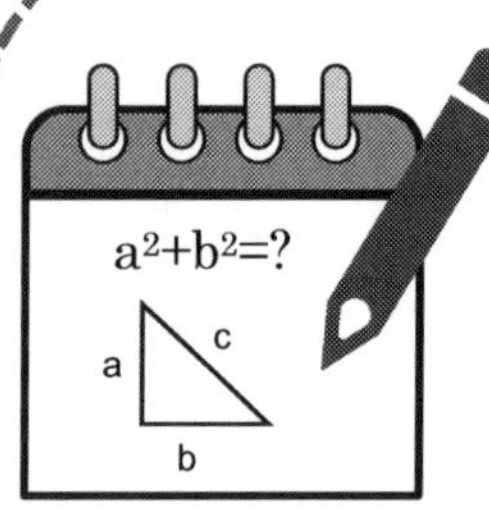

This study guide will:

- ☑ Explain the format of the Praxis Core Math test.
- ☑ Describe specific test-taking strategies that you can use on the test.
- ☑ Provide Praxis Math test-taking tips.
- ☑ Review all Praxis Math concepts and topics you will be tested on.
- ☑ Help you identify the areas in which you need to concentrate your study time.
- ☑ Offer exercises that help you develop the basic math skills you will learn in each section.
- ☑ Give **2 realistic and full-length practice tests** (featuring new question types) with detailed answers to help you measure your exam readiness and build confidence.

This resource contains everything you will ever need to succeed on the Praxis Core Math test. You'll get in-depth instructions on every math topic as well as tips and techniques on how to answer each question type. You'll also get plenty of practice questions to boost your test-taking confidence.

In addition, in the following pages you'll find:

- **How to Use This Book Effectively** – This section provides you with step-by-step instructions on how to get the most out of this comprehensive study guide.
- **How to study for the Praxis Core Math Test** – A six-step study program has been developed to help you make the best use of this book and prepare for your Praxis Math test. Here you'll find tips and strategies to guide your study program and help you understand Praxis Math and how to ace the test.

- **Praxis Math Review** – Learn everything you need to know about the Praxis Core Math test.
- **Praxis Math Test-Taking Strategies** – Learn how to effectively put these recommended test-taking techniques into use for improving your Praxis Core Math score.
- **Test Day Tips** – Review these tips to make sure you will do your best when the big day comes.

Effortless Math's Praxis Core Online Center

Effortless Math Online Praxis Core Center offers a complete study program, including the following:

- ✓ Step-by-step instructions on how to prepare for the Praxis Core Math test
- ✓ Numerous Praxis Core Math worksheets to help you measure your math skills
- ✓ Complete list of Praxis Math formulas
- ✓ Video lessons for all Praxis Core Math topics
- ✓ Full-length Praxis Math practice tests
- ✓ And much more…

No Registration Required.

Visit **EffortlessMath.com/Praxis** to find your online Praxis Core Math resources.

How to Use This Book Effectively

Look no further when you need a study guide to improve your math skills to succeed on the math portion of the Praxis Core test. Each chapter of this comprehensive guide to the Praxis Core Math will provide you with the knowledge, tools, and understanding needed for every topic covered on the test.

It's imperative that you understand each topic before moving onto another one, as that's the way to guarantee your success. Each chapter provides you with examples and a step-by-step guide of every concept to better understand the content that will be on the test. To get the best possible results from this book:

- **Begin studying long before your test date**. This provides you ample time to learn the different math concepts. The earlier you begin studying for the test, the sharper your skills will be. Do not procrastinate! Provide yourself with plenty of time to learn the concepts and feel comfortable that you understand them when your test date arrives.

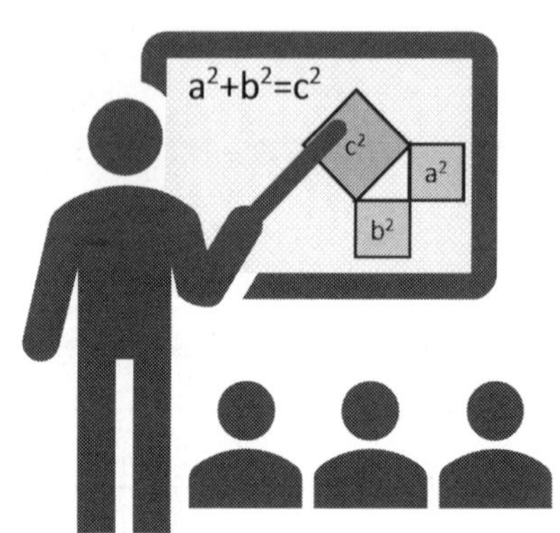

- **Practice consistently**. Study Praxis Core Math concepts at least 20 to 30 minutes a day. Remember, slow and steady wins the race, which can be applied to preparing for the Praxis Core Math test. Instead of cramming to tackle everything at once, be patient and learn the math topics in short bursts.
- Whenever you get a math problem wrong, **mark it off, and review it later** to make sure you understand the concept.
- Start each session by **looking over the previous material.**
- Once you've reviewed the book's lessons, **take a practice test at the back of the book** to gauge your level of readiness. Then, review your results. Read detailed answers and solutions for each question you missed.
- **Take another practice test** to get an idea of how ready you are to take the actual exam. Taking the practice tests will give you the confidence you need on test day. Simulate the Praxis Core testing environment by sitting in a quiet room free from distraction. Make sure to clock yourself with a timer.

How to Study for the Praxis Core Math Test

Studying for the Praxis Core Math test can be a really daunting and boring task. What's the best way to go about it? Is there a certain study method that works better than others? Well, studying for the Praxis Math can be done effectively. The following six-step program has been designed to make preparing for the Praxis Math test more efficient and less overwhelming.

Step 1 - Create a study plan
Step 2 - Choose your study resources
Step 3 - Review, Learn, Practice
Step 4 - Learn and practice test-taking strategies
Step 5 - Learn the Praxis Core Test format and take practice tests
Step 6 - Analyze your performance

STEP 1: Create a Study Plan

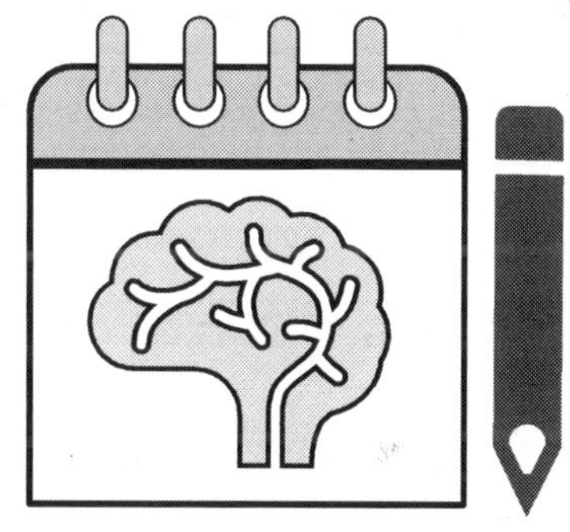

It's always easier to get things done when you have a plan. Creating a study plan for the Praxis Core Math test can help you to stay on track with your studies. It's important to sit down and prepare a study plan with what works with your life, work, and any other obligations you may have. Devote enough time each day to studying. It's also a great idea to break down each section of the exam into blocks and study one concept at a time.

It's important to understand that there is no "right" way to create a study plan. Your study plan will be personalized based on your specific needs and learning style.

Follow these guidelines to create an effective study plan for your Praxis Math test:

★ **Analyze your learning style and study habits** – Everyone has a different learning style. It is essential to embrace your individuality and the unique way you learn. Think about what works and what doesn't work for you. Do you prefer Praxis Math prep books or a combination of textbooks and video lessons? Does it work better for you if you study every night for thirty minutes or is it more effective to study in the morning before going to work?

★ **Evaluate your schedule** – Review your current schedule and find out how much time you can consistently devote to Praxis Core Math study.

★ **Develop a schedule** – Now it's time to add your study schedule to your calendar like any other obligation. Schedule time for study, practice, and review. Plan out which topic you will study on which day to ensure that you're devoting enough time to each concept. Develop a study plan that is mindful, realistic, and flexible.

★ **Stick to your schedule** – A study plan is only effective when it is followed consistently. You should try to develop a study plan that you can follow for the length of your study program.

★ **Evaluate your study plan and adjust as needed** – Sometimes you need to adjust your plan when you have new commitments. Check in with yourself regularly to make sure that you're not falling behind in your study plan. Remember, the most important thing is sticking to your plan. Your study plan is all about helping you be more productive. If you find that your study plan is not as effective as you want, don't get discouraged. It's okay to make changes as you figure out what works best for you.

STEP 2: Choose Your Study Resources

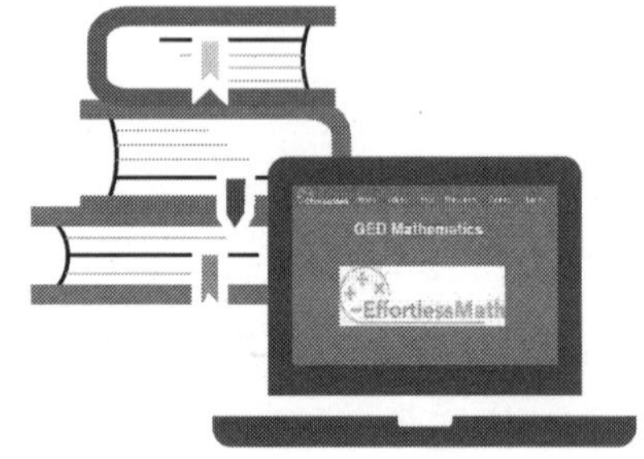

There are numerous textbooks and online resources available for the Praxis Core Math test, and it may not be clear where to begin. Don't worry! This study guide provides everything you need to fully prepare for your Praxis Core Math test. In addition to the book content, you can also use Effortless Math's online resources. (video lessons, worksheets, formulas, etc.) On each page, there is a link (and a QR code) to an online webpage which provides a comprehensive review of the topic, step-by-step instruction, video tutorial, and numerous examples and exercises to help you fully understand the concept.

You can also visit EffortlessMath.com/Praxis to find your online Praxis Core Math resources.

Step 3: Review, Learn, Practice

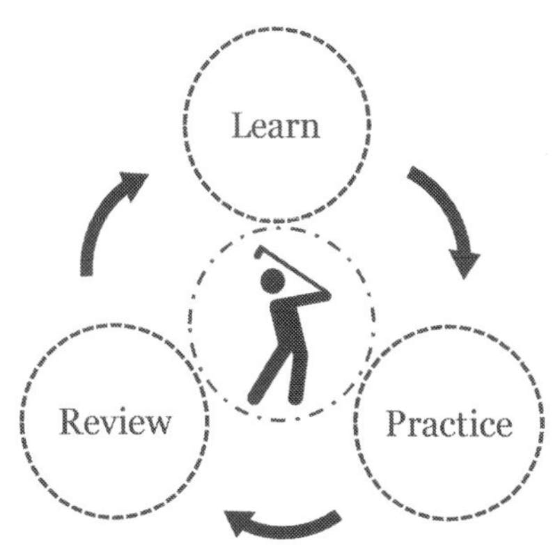

This Praxis Core Math study guide breaks down each subject into specific skills or content areas. For instance, the percent concept is separated into different topics–percent calculation, percent increase and decrease, percent problems, etc. Use this study guide and Effortless Math online Praxis Core center to help you go over all key math concepts and topics on the Praxis Core Math test.

As you read each topic, take notes or highlight the concepts you would like to go over again in the future. If you're unfamiliar with a topic or something is difficult for you, use the link (or the QR code) at the bottom of the page to find the webpage that provides more instruction about that topic. For each math topic, plenty of instructions, step-by-step guides, and examples are provided to ensure you get a good grasp of the material.

Quickly review the topics you do understand to get a brush-up of the material. Be sure to do the practice questions provided at the end of every chapter to measure your understanding of the concepts.

Step 4: Learn and Practice Test-taking Strategies

In the following sections, you will find important test-taking strategies and tips that can help you earn extra points. You'll learn how to think strategically and when to guess if you don't know the answer to a question. Using Praxis Core Math test-taking strategies and tips can help you raise your score and do well on the test. Apply test taking strategies on the practice tests to help you boost your confidence.

Step 5: Learn the Praxis Core Test Format and Take Practice Tests

The Praxis Core *Test Review* section provides information about the structure of the Praxis test. Read this section to learn more about the Praxis test structure, different test sections, the number of questions in each section, and the section time limits. When you have a prior understanding of the test format and different types of Praxis Math questions, you'll feel more confident when you take the actual exam.

Once you have read through the instructions and lessons and feel like you are ready to go – take advantage of both of the full-length Praxis Core Math practice tests available in this study guide. Use the practice tests to sharpen your skills and build confidence.

The Praxis Math practice tests offered at the end of the book are formatted similarly to the actual Praxis Math test. When you take each practice test, try to simulate actual testing conditions. To take the practice tests, sit in a quiet space, time yourself, and work through as many of the questions as time allows. The practice tests are followed by detailed answer explanations to help you find your weak areas, learn from your mistakes, and raise your Praxis Math score.

Step 6: Analyze Your Performance

After taking the practice tests, look over the answer keys and explanations to learn which questions you answered correctly and which you did not. Never be discouraged if you make a few mistakes. See them as a learning opportunity. This will highlight your strengths and weaknesses.

You can use the results to determine if you need additional practice or if you are ready to take the actual Praxis Core Math test.

Looking for more?

Visit EffortlessMath.com/Praxis to find hundreds of Praxis Core Math worksheets, video tutorials, practice tests, Praxis Core Math formulas, and much more.

Or scan this QR code.

No Registration Required.

Praxis Core Test Review

The Praxis Core Academic Skills for Educators is a standardized test used for admissions to teacher preparation programs in the United States. In essence, it is a broad and quick assessment of students' academic abilities. The exam was designed and is administered by the Educational Testing Service (ETS) and scoring well on this exam is vital to being accepted for admission into teacher preparation programs.

The Praxis Core test covers three topics.

- Math
- Reading
- Writing

The Praxis Core Math (5733) is comprised of 56 multiple choice and numeric entry questions and test takers have 90 minutes to answer the questions. You will be able to use a basic on-screen calculator on Praxis Core Math test.

Praxis Core Mathematics cover the following topics:

- Number and Quantity (36%)
- Algebra and Functions (20%)
- Geometry (12%)
- Data Interpretation, Statistics and Probability (32%)

Praxis Core Math Test-Taking Strategies

Here are some test-taking strategies that you can use to maximize your performance and results on the Praxis Core Math test.

#1: Use This Approach To Answer Every Praxis Core Math Question

- Review the question to identify keywords and important information.
- Translate the keywords into math operations so you can solve the problem.
- Review the answer choices. What are the differences between answer choices?
- Draw or label a diagram if needed.
- Try to find patterns.
- Find the right method to answer the question. Use straightforward math, plug in numbers, or test the answer choices (backsolving).
- Double-check your work.

#2: Use Educated Guessing

This approach is applicable to the problems you understand to some degree but cannot solve using straightforward math. In such cases, try to filter out as many answer choices as possible before picking an answer. In cases where you don't have a clue about what a certain problem entails, don't waste any time trying to eliminate answer choices. Just choose one randomly before moving onto the next question.

As you can ascertain, direct solutions are the most optimal approach. Carefully read through the question, determine what the solution is using the math you have learned before, then coordinate the answer with one of the choices available to you. Are you stumped? Make your best guess, then move on.

Don't leave any fields empty! Even if you're unable to work out a problem, strive to answer it. Take a guess if you have to. You will not lose points by getting an answer wrong, though you may gain a point by getting it correct!

#3 : BALLPARK

A ballpark answer is a rough approximation. When we become overwhelmed by calculations and figures, we end up making silly mistakes. A decimal that is moved by one unit can change an answer from right to wrong, regardless of the number of steps that you went through to get it. That's where ballparking can play a big part.

If you think you know what the correct answer may be (even if it's just a ballpark answer), you'll usually have the ability to eliminate a couple of choices. While answer choices are usually based on the average student error and/or values that are closely tied, you will still be able to weed out choices that are way far afield. Try to find answers that aren't in the proverbial ballpark when you're looking for a wrong answer on a multiple-choice question. This is an optimal approach to eliminating answers to a problem.

#4 : BACKSOLVING

A majority of questions on the Praxis Core Math test will be in multiple-choice format. Many test-takers prefer multiple-choice questions, as at least the answer is right there. You'll typically have five answers to pick from. You simply need to figure out which one is correct. Usually, the best way to go about doing so is "backsolving."

As mentioned earlier, direct solutions are the most optimal approach to answering a question. Carefully read through a problem, calculate a solution, then correspond the answer with one of the choices displayed in front of you. If you can't calculate a solution, your next best approach involves "backsolving."

When backsolving a problem, contrast one of your answer options against the problem you are asked, then see which of them is most relevant. More often than not, answer choices are listed in ascending or descending order. In such cases, try out the choices B or C. If it's not correct, you can go either down or up from there.

#5 : Plugging In Numbers

"Plugging in numbers" is a strategy that can be applied to a wide range of different math problems on the Praxis Core Math test. This approach is typically used to simplify a challenging question so that it is more understandable. By using the strategy carefully, you can find the answer without too much trouble.

The concept is fairly straightforward–replace unknown variables in a problem with certain values. When selecting a number, consider the following:

- Choose a number that's basic (just not too basic). Generally, you should avoid choosing 1 (or even 0). A decent choice is 2.
- Try not to choose a number that is displayed in the problem.
- Make sure you keep your numbers different if you need to choose at least two of them.
- More often than not, choosing numbers merely lets you filter out some of your answer choices. As such, don't just go with the first choice that gives you the right answer.
- If several answers seem correct, then you'll need to choose another value and try again. This time, though, you'll just need to check choices that haven't been eliminated yet.
- If your question contains fractions, then a potential right answer may involve either an LCD (least common denominator) or an LCD multiple.
- 100 is the number you should choose when you are dealing with problems involving percentages.

Praxis Core Mathematics Test – Daytime Tips

After practicing and reviewing all the math concepts you've been taught, and taking some Praxis Core mathematics practice tests, you'll be prepared for test day. Consider the following tips to be extra-ready come test time.

Before Your Test

What to do the night before:

- **Relax!** One day before your test, study lightly or skip studying altogether. You shouldn't attempt to learn something new, either. There are plenty of reasons why studying the evening before a big test can work against you. Put it this way–a marathoner wouldn't go out for a sprint before the day of a big race. Mental marathoners–such as yourself–should not study for any more than one hour 24 hours before a Praxis Core test. That's because your brain requires some rest to be at its best. The night before your exam, spend some time with family or friends, or read a book.
- **Avoid bright screens** - You'll have to get some good shuteye the night before your test. Bright screens (such as the ones coming from your laptop, TV, or mobile device) should be avoided altogether. Staring at such a screen will keep your brain up, making it hard to drift asleep at a reasonable hour.
- **Make sure your dinner is healthy** - The meal that you have for dinner should be nutritious. Be sure to drink plenty of water as well. Load up on your complex carbohydrates, much like a marathon runner would do. Pasta, rice, and potatoes are ideal options here, as are vegetables and protein sources.
- **Get your bag ready for test day** - The night prior to your test, pack your bag with your stationery, admissions pass, ID, and any other gear that you need. Keep the bag right by your front door.
- **Make plans to reach the testing site** - Before going to sleep, ensure that you understand precisely how you will arrive at the site of the test. If parking is something you'll have to find first, plan for it. If you're dependent on public transit, then review the schedule. You should also make sure that the train/bus/subway/streetcar you use will be running. Find out about road closures as well. If a parent or friend is accompanying you, ensure that they understand what steps they have to take as well.

The Day of the Test

- **Get up reasonably early, but not too early.**
- **Have breakfast** - Breakfast improves your concentration, memory, and mood. As such, make sure the breakfast that you eat in the morning is healthy. The last thing you want to be is distracted by a grumbling tummy. If it's not your own stomach making those noises, another test taker close to you might be instead. Prevent discomfort or embarrassment by consuming a healthy breakfast. Bring a snack with you if you think you'll need it.
- **Follow your daily routine** - Do you watch Good Morning America each morning while getting ready for the day? Don't break your usual habits on the day of the test. Likewise, if coffee isn't something you drink in the morning, then don't take up the habit hours before your test. Routine consistency lets you concentrate on the main objective–doing the best you can on your test.
- **Wear layers** - Dress yourself up in comfortable layers. You should be ready for any kind of internal temperature. If it gets too warm during the test, take a layer off.
- **Get there on time** - The last thing you want to do is get to the test site late. Rather, you should be there 45 minutes prior to the start of the test. Upon your arrival, try not to hang out with anybody who is nervous. Any anxious energy they exhibit shouldn't influence you.
- **Leave the books at home** - No books should be brought to the test site. If you start developing anxiety before the test, books could encourage you to do some last-minute studying, which will only hinder you. Keep the books far away–better yet, leave them at home.
- **Make your voice heard** - If something is off, speak to a proctor. If medical attention is needed or if you'll require anything, consult the proctor prior to the start of the test. Any doubts you have should be clarified. You should be entering the test site with a state of mind that is completely clear.

- **Have faith in yourself** - When you feel confident, you will be able to perform at your best. When you are waiting for the test to begin, envision yourself receiving an outstanding result. Try to see yourself as someone who knows all the answers, no matter what the questions are. A lot of athletes tend to use this technique–particularly before a big competition. Your expectations will be reflected by your performance.

During your test

- **Be calm and breathe deeply** - You need to relax before the test, and some deep breathing will go a long way to help you do that. Be confident and calm. You got this. Everybody feels a little stressed out just before an evaluation of any kind is set to begin. Learn some effective breathing exercises. Spend a minute meditating before the test starts. Filter out any negative thoughts you have. Exhibit confidence when having such thoughts.

- **Concentrate on the test** - Refrain from comparing yourself to anyone else. You shouldn't be distracted by the people near you or random noise. Concentrate exclusively on the test. If you find yourself irritated by surrounding noises, earplugs can be used to block sounds off close to you. Don't forget–the test is going to last several hours if you're taking more than one subject of the test. Some of that time will be dedicated to brief sections. Concentrate on the specific section you are working on during a particular moment. Do not let your mind wander off to upcoming or previous sections.

- **Skip challenging questions** - Optimize your time when taking the test. Lingering on a single question for too long will work against you. If you don't know what the answer is to a certain question, use your best guess, and mark the question so you can review it later on. There is no need to spend time attempting to solve something you aren't sure about. That time would be better served handling the questions you can actually answer well. You will not be penalized for getting the wrong answer on a test like this.

- **Try to answer each question individually** - Focus only on the question you are working on. Use one of the test-taking strategies to solve the problem. If you aren't able to come up with an answer, don't get frustrated. Simply skip that question, then move onto the next one.

- **Don't forget to breathe!** Whenever you notice your mind wandering, your stress levels boosting, or frustration brewing, take a thirty-second break. Shut your eyes, drop your pencil, breathe deeply, and let your shoulders relax. You will end up being more productive when you allow yourself to relax for a moment.
- **Review your answer.** If you still have time at the end of the test, don't waste it. Go back and check over your answers. It is worth going through the test from start to finish to ensure that you didn't make a sloppy mistake somewhere.
- **Optimize your breaks** - When break time comes, use the restroom, have a snack, and reactivate your energy for the subsequent section. Doing some stretches can help stimulate your blood flow.

After your test

- **Take it easy** - You will need to set some time aside to relax and decompress once the test has concluded. There is no need to stress yourself out about what you could've said, or what you may have done wrong. At this point, there's nothing you can do about it. Your energy and time would be better spent on something that will bring you happiness for the remainder of your day.
- **Redoing the test** - Did you pass the test? Congratulations! Your hard work paid off!

  If you have failed your test, though, don't worry! The test can be retaken. In such cases, you will need to follow the retake policy. You also need to re-register to take the exam again.

Contents

Chapter 1 Fractions and Mixed Numbers

Math topics that you'll learn in this chapter:

- ☑ Simplifying Fractions
- ☑ Adding and Subtracting Fractions
- ☑ Multiplying and Dividing Fractions
- ☑ Adding Mixed Numbers
- ☑ Subtracting Mixed Numbers
- ☑ Multiplying Mixed Numbers
- ☑ Dividing Mixed Numbers

Simplifying Fractions

- A fraction contains two numbers separated by a bar between them. The bottom number, called the denominator, is the total number of equally divided portions in one whole. The top number, called the numerator, is how many portions you have. And the bar represents the operation of division.
- Simplifying a fraction means reducing it to the lowest terms. To simplify a fraction, evenly divide both the top and bottom of the fraction by $2, 3, 5, 7$, etc.
- Continue until you can't go any further.

Examples:

Example 1. Simplify $\frac{18}{30}$

Solution: To simplify $\frac{18}{30}$, find a number that both 18 and 30 are divisible by. Both are divisible by 6. Then: $\frac{18}{30} = \frac{18 \div 6}{30 \div 6} = \frac{3}{5}$

Example 2. Simplify $\frac{32}{80}$

Solution: To simplify $\frac{32}{80}$, find a number that both 32 and 80 are divisible by. Both are divisible by 8 and 16. Then: $\frac{32}{80} = \frac{32 \div 8}{80 \div 8} = \frac{4}{10}$, 4 and 10 are divisible by 2, then: $\frac{4}{10} = \frac{2}{5}$ or $\frac{32}{80} = \frac{32 \div 16}{80 \div 16} = \frac{2}{5}$

Practices:

Simplify each fraction.

1) $\frac{4}{14} =$
2) $\frac{9}{24} =$
3) $\frac{6}{10} =$
4) $\frac{7}{28} =$
5) $\frac{25}{200} =$
6) $\frac{3}{9} =$

Adding and Subtracting Fractions

- For "like" fractions (fractions with the same denominator), add or subtract the numerators (top numbers) and write the answer over the common denominator (bottom numbers).
- Adding and Subtracting fractions with the same denominator:

$$\frac{a}{b}+\frac{c}{b}=\frac{a+c}{b} \qquad \frac{a}{b}-\frac{c}{b}=\frac{a-c}{b}$$

- Find equivalent fractions with the same denominator before you can add or subtract fractions with different denominators.
- Adding and Subtracting fractions with different denominators:

$$\frac{a}{b}+\frac{c}{d}=\frac{ad+bc}{bd} \qquad \frac{a}{b}-\frac{c}{d}=\frac{ad-bc}{bd}$$

Examples:

Example 1. Find the sum. $\frac{2}{3}+\frac{1}{2}=$

Solution: These two fractions are "unlike" fractions. (they have different denominators). Use this formula: $\frac{a}{b}+\frac{c}{d}=\frac{ad+cb}{bd}$. Then: $\frac{2}{3}+\frac{1}{2}=\frac{(2)(2)+(3)(1)}{3\times2}=\frac{4+3}{6}=\frac{7}{6}$

Example 2. Find the difference. $\frac{3}{5}-\frac{2}{7}=$

Solution: For "unlike" fractions, find equivalent fractions with the same denominator before you can add or subtract fractions with different denominators. Use this formula: $\frac{a}{b}-\frac{c}{d}=\frac{ad-bc}{bd}$

$\frac{3}{5}-\frac{2}{7}=\frac{(3)(7)-(2)(5)}{5\times7}=\frac{21-10}{35}=\frac{11}{35}$

Practices:

Find the sum or difference.

1) $\frac{2}{3}+\frac{3}{4}=$
2) $\frac{1}{2}-\frac{1}{5}=$
3) $\frac{2}{7}+\frac{1}{2}=$
4) $\frac{1}{3}-\frac{2}{7}=$
5) $\frac{1}{2}-\frac{1}{4}=$
6) $\frac{3}{5}+\frac{3}{3}=$

bit.ly/3x5jfwe

Find more at

Multiplying and Dividing Fractions

- **Multiplying fractions:** multiply the top numbers and multiply the bottom numbers. Simplify if necessary. $\frac{a}{b}\times\frac{c}{d}=\frac{a\times c}{b\times d}$
- **Dividing fractions:** Keep, Change, Flip
- Keep the first fraction, change the division sign to multiplication, and flip the numerator and denominator of the second fraction. Then, solve!

$$\frac{a}{b}\div\frac{c}{d}=\frac{a}{b}\times\frac{d}{c}=\frac{a\times d}{b\times c}$$

Examples:

Example 1. Multiply. $\frac{2}{3}\times\frac{3}{5}=$

Solution: Multiply the top numbers and multiply the bottom numbers.
$\frac{2}{3}\times\frac{3}{5}=\frac{2\times3}{3\times5}=\frac{6}{15}$, now, simplify: $\frac{6}{15}=\frac{6\div3}{15\div3}=\frac{2}{5}$

Example 2. Solve. $\frac{3}{4}\div\frac{2}{5}=$

Solution: Keep the first fraction, change the division sign to multiplication, and flip the numerator and denominator of the second fraction.
Then: $\frac{3}{4}\div\frac{2}{5}=\frac{3}{4}\times\frac{5}{2}=\frac{3\times5}{4\times2}=\frac{15}{8}$

Practices:

Find the answers.

1) $\frac{3}{10}\times\frac{1}{2}=$

2) $\frac{1}{5}\div\frac{5}{6}=$

3) $\frac{3}{4}\times\frac{1}{7}=$

4) $\frac{1}{6}\div\frac{2}{5}=$

5) $\frac{2}{3}\times\frac{3}{4}=$

6) $\frac{3}{7}\div\frac{3}{4}=$

Adding Mixed Numbers

Use the following steps for adding mixed numbers:

- Add whole numbers of the mixed numbers.
- Add the fractions of the mixed numbers.
- Find the Least Common Denominator (LCD) if necessary.
- Add whole numbers and fractions.
- Write your answer in the lowest terms.

Examples:

Example 1. Add mixed numbers. $2\frac{1}{2} + 1\frac{2}{3} =$

Solution: Let's rewriting our equation with parts separated, $2\frac{1}{2} + 1\frac{2}{3} = 2 + \frac{1}{2} + 1 + \frac{2}{3}$.

Now, add whole number parts: $2 + 1 = 3$

Add the fraction parts $\frac{1}{2} + \frac{2}{3}$. Rewrite to solve with the equivalent fractions. $\frac{1}{2} + \frac{2}{3} = \frac{3}{6} + \frac{4}{6} = \frac{7}{6}$. The answer is an improper fraction (numerator is bigger than denominator). Convert the improper fraction into a mixed number: $\frac{7}{6} = 1\frac{1}{6}$. Now, combine the whole and fraction parts: $3 + 1\frac{1}{6} = 4\frac{1}{6}$

Example 2. Find the sum. $1\frac{3}{4} + 2\frac{1}{2} =$

Solution: Rewriting our equation with parts separated, $1 + \frac{3}{4} + 2 + \frac{1}{2}$. Add the whole number parts:

$1 + 2 = 3$. Add the fraction parts: $\frac{3}{4} + \frac{1}{2} = \frac{3}{4} + \frac{2}{4} = \frac{5}{4}$

Convert the improper fraction into a mixed number: $\frac{5}{4} = 1\frac{1}{4}$.

Now, combine the whole and fraction parts: $3 + 1\frac{1}{4} = 4\frac{1}{4}$

Practices:

Find the sum.

1) $5\frac{2}{3} + 2\frac{1}{2} =$

2) $3\frac{1}{2} + 4\frac{1}{2} =$

3) $2\frac{3}{8} + 2\frac{1}{8} =$

4) $4\frac{1}{7} + 6\frac{1}{14} =$

5) $7\frac{1}{5} + 1\frac{4}{15} =$

6) $3\frac{1}{3} + 3\frac{3}{4} =$

Subtracting Mixed Numbers

Use these steps for subtracting mixed numbers.

- Convert mixed numbers into improper fractions. $a\frac{c}{b}=\frac{ab+c}{b}$
- Find equivalent fractions with the same denominator for unlike fractions. (fractions with different denominators)
- Subtract the second fraction from the first one. $\frac{a}{b}-\frac{c}{d}=\frac{ad-bc}{bd}$
- Write your answer in the lowest terms.
- If the answer is an improper fraction, convert it into a mixed number.

Examples:

Example 1. Subtract. $2\frac{1}{3}-1\frac{1}{2}=$

Solution: Convert mixed numbers into fractions: $2\frac{1}{3}=\frac{2\times3+1}{3}=\frac{7}{3}$ and $1\frac{1}{2}=\frac{1\times2+1}{2}=\frac{3}{2}$

These two fractions are "unlike" fractions. (they have different denominators). Find equivalent fractions with the same denominator. Use this formula: $\frac{a}{b}-\frac{c}{d}=\frac{ad-bc}{bd}$

$\frac{7}{3}-\frac{3}{2}=\frac{(7)(2)-(3)(3)}{3\times2}=\frac{14-9}{6}=\frac{5}{6}$

Example 2. Find the difference. $3\frac{4}{7}-2\frac{3}{4}=$

Solution: Convert mixed numbers into fractions: $3\frac{4}{7}=\frac{3\times7+4}{7}=\frac{25}{7}$ and $2\frac{3}{4}=\frac{2\times4+3}{4}=\frac{11}{4}$. Then: $3\frac{4}{7}-2\frac{3}{4}=\frac{25}{7}-\frac{11}{4}=\frac{(25)(4)-(11)(7)}{7\times4}=\frac{23}{28}$

Practices:

Find the difference.

1) $2\frac{2}{3}-1\frac{1}{3}=$	3) $2\frac{3}{8}-2\frac{1}{4}=$	5) $5\frac{1}{2}-2\frac{3}{10}=$
2) $6\frac{1}{2}-3\frac{1}{2}=$	4) $5\frac{1}{3}-3\frac{5}{6}=$	6) $10\frac{2}{3}-4\frac{1}{4}=$

Find more at bit.ly/3aD3KDG

Multiplying Mixed Numbers

Use the following steps for multiplying mixed numbers:

- Convert the mixed numbers into fractions. $a\frac{c}{b} = a + \frac{c}{b} = \frac{ab+c}{b}$
- Multiply fractions. $\frac{a}{b} \times \frac{c}{d} = \frac{a \times c}{b \times d}$
- Write your answer in the lowest terms.
- If the answer is an improper fraction (numerator is bigger than denominator), convert it into a mixed number.

Examples:

Example 1. Multiply. $4\frac{1}{2} \times 2\frac{2}{5} =$

Solution: Convert mixed numbers into fractions, $4\frac{1}{2} = \frac{4\times2+1}{2} = \frac{9}{2}$ and $2\frac{2}{5} = \frac{2\times5+2}{5} = \frac{12}{5}$. Apply the fractions rule for multiplication: $\frac{9}{2} \times \frac{12}{5} = \frac{9\times12}{2\times5} = \frac{108}{10} = \frac{54}{5}$
The answer is an improper fraction. Convert it into a mixed number. $\frac{54}{5} = 10\frac{4}{5}$

Example 2. Multiply. $3\frac{2}{3} \times 2\frac{5}{6} =$

Solution: Converting mixed numbers into fractions, $3\frac{2}{3} \times 2\frac{5}{6} = \frac{11}{3} \times \frac{17}{6}$
Apply the fractions rule for multiplication: $\frac{11}{3} \times \frac{17}{6} = \frac{11\times17}{3\times6} = \frac{187}{18} = 10\frac{7}{18}$

Practices:

Find the product.

1) $3\frac{2}{3} \times 2\frac{2}{5} =$
2) $5\frac{1}{2} \times 3\frac{1}{3} =$
3) $4\frac{1}{4} \times 1\frac{3}{5} =$
4) $10\frac{1}{8} \times 2\frac{2}{9} =$
5) $7\frac{1}{5} \times 2\frac{3}{4} =$
6) $7\frac{1}{3} \times 1\frac{1}{11} =$

Dividing Mixed Numbers

Use the following steps for dividing mixed numbers:

- Convert the mixed numbers into fractions. $a\frac{c}{b} = a + \frac{c}{b} = \frac{ab+c}{b}$
- Divide fractions: Keep, Change, Flip: Keep the first fraction, change the division sign to multiplication, and flip the numerator and denominator of the second fraction. Then, solve! $\frac{a}{b} \div \frac{c}{d} = \frac{a}{b} \times \frac{d}{c} = \frac{a \times d}{b \times c}$
- Write your answer in the lowest terms.
- If the answer is an improper fraction (numerator is bigger than denominator), convert it into a mixed number.

Examples:

Example 1. Solve. $2\frac{1}{3} \div 1\frac{1}{2}$

Solution: Convert mixed numbers into fractions: $2\frac{1}{3} = \frac{2\times3+1}{3} = \frac{7}{3}$ and $1\frac{1}{2} = \frac{1\times2+1}{2} = \frac{3}{2}$ Keep, Change, Flip: $\frac{7}{3} \div \frac{3}{2} = \frac{7}{3} \times \frac{2}{3} = \frac{7\times2}{3\times3} = \frac{14}{9}$. The answer is an improper fraction. Convert it into a mixed number: $\frac{14}{9} = 1\frac{5}{9}$

Example 2. Solve. $3\frac{3}{4} \div 2\frac{2}{5}$

Solution: Convert mixed numbers to fractions, then solve:
$3\frac{3}{4} \div 2\frac{2}{5} = \frac{15}{4} \div \frac{12}{5} = \frac{15}{4} \times \frac{5}{12} = \frac{75}{48} = 1\frac{9}{16}$

Practices:

Find the quotient.

1) $2\frac{2}{3} \div 3\frac{2}{3} =$

2) $10\frac{1}{4} \div 1\frac{1}{2} =$

3) $1\frac{3}{7} \div 2\frac{2}{7} =$

4) $4\frac{1}{6} \div 3\frac{1}{3} =$

5) $5\frac{1}{5} \div 2\frac{1}{10} =$

6) $1\frac{3}{8} \div 2\frac{3}{4} =$

Chapter 1: Answers

Simplifying Fractions

1) $\frac{2}{7}$ (To simplify $\frac{4}{14}$, find a number that both 4 and 14 are divisible by. Both are divisible by 2. Then: $\frac{4}{14} = \frac{4 \div 2}{14 \div 2} = \frac{2}{7}$)

2) $\frac{3}{8}$ (To simplify $\frac{9}{24}$, find a number that both 9 and 24 are divisible by. Both are divisible by 3. Then: $\frac{9}{24} = \frac{9 \div 3}{24 \div 3} = \frac{3}{8}$)

3) $\frac{3}{5}$ (To simplify $\frac{6}{10}$, find a number that both 6 and 10 are divisible by. Both are divisible by 2. Then: $\frac{6}{10} = \frac{6 \div 2}{10 \div 2} = \frac{3}{5}$)

4) $\frac{1}{4}$ (To simplify $\frac{7}{28}$, find a number that both 7 and 28 are divisible by. Both are divisible by 7. Then: $\frac{7}{28} = \frac{7 \div 7}{28 \div 7} = \frac{1}{4}$)

5) $\frac{1}{8}$ (To simplify $\frac{25}{200}$, find a number that both 25 and 200 are divisible by. Both are divisible by 25. Then: $\frac{25}{200} = \frac{25 \div 25}{200 \div 25} = \frac{1}{8}$)

6) $\frac{1}{3}$ (To simplify $\frac{3}{9}$, find a number that both 3 and 9 are divisible by. Both are divisible by 3. Then: $\frac{3}{9} = \frac{3 \div 3}{9 \div 3} = \frac{1}{3}$)

Adding and Subtracting Fractions

1) $\frac{17}{12}$ (These two fractions are "unlike" fractions. (they have different denominators). Use this formula: $\frac{a}{b} + \frac{c}{d} = \frac{ad+cb}{bd}$. Then: $\frac{2}{3} + \frac{3}{4} = \frac{(2)(4)+(3)(3)}{3 \times 4} = \frac{8+9}{12} = \frac{17}{12}$)

2) $\frac{3}{10}$ (For "unlike" fractions, find equivalent fractions with the same denominator before you can add or subtract fractions with different denominators. Use this formula: $\frac{a}{b} - \frac{c}{d} = \frac{ad-bc}{bd} = \frac{1}{2} - \frac{1}{5} = \frac{(1)(5)-(1)(2)}{2 \times 5} = \frac{5-2}{10} = \frac{3}{10}$)

3) $\frac{11}{14}$ (These two fractions are "unlike" fractions. (they have different denominators). Use this formula: $\frac{a}{b}+\frac{c}{d}=\frac{ad+cb}{bd}$. Then: $\frac{2}{7}+\frac{1}{2}=\frac{(2)(2)+(1)(7)}{2\times7}=\frac{4+7}{14}=\frac{11}{14}$)

4) $\frac{1}{21}$ (For "unlike" fractions, find equivalent fractions with the same denominator before you can add or subtract fractions with different denominators. Use this formula: $\frac{a}{b}-\frac{c}{d}=\frac{ad-bc}{bd}=\frac{1}{3}-\frac{2}{7}=\frac{(1)(7)-(2)(3)}{3\times7}=\frac{7-6}{21}=\frac{1}{21}$)

5) $\frac{1}{4}$ (For "unlike" fractions, find equivalent fractions with the same denominator before you can add or subtract fractions with different denominators. Use this formula: $\frac{a}{b}-\frac{c}{d}=\frac{ad-cb}{bd}=\frac{1}{2}-\frac{1}{4}=\frac{(1)(2)-(1)}{4}=\frac{2-1}{4}=\frac{1}{4}$)

6) $\frac{24}{15}$ (These two fractions are "unlike" fractions. (they have different denominators). Use this formula: $\frac{a}{b}+\frac{c}{d}=\frac{ad+cb}{bd}$. Then: $\frac{3}{5}+\frac{3}{3}=\frac{(3)(3)+(3)(5)}{5\times3}=\frac{9+15}{15}=\frac{24}{15}$)

Multiplying and Dividing Fractions

1) $\frac{3}{20}$ (Multiply the top numbers and multiply the bottom numbers.

$\frac{3}{10}\times\frac{1}{2}=\frac{3\times1}{10\times2}=\frac{3}{20}$)

2) $\frac{6}{25}$ (Keep the first fraction, change the division sign to multiplication, and flip the numerator and denominator of the second fraction. Then: $\frac{1}{5}\div\frac{5}{6}=\frac{1}{5}\times\frac{6}{5}=\frac{1\times6}{5\times5}=\frac{6}{25}$)

3) $\frac{3}{28}$ (Multiply the top numbers and multiply the bottom numbers.

$\frac{3}{4}\times\frac{1}{7}=\frac{3\times1}{4\times7}=\frac{3}{28}$)

4) $\frac{5}{12}$ (Keep the first fraction, change the division sign to multiplication, and flip the numerator and denominator of the second fraction. Then: $\frac{1}{6} \div \frac{2}{5} = \frac{1}{6} \times \frac{5}{2} = \frac{1 \times 5}{6 \times 2} = \frac{5}{12}$)

5) $\frac{1}{2}$ (Multiply the top numbers and multiply the bottom numbers.

$\frac{2}{3} \times \frac{3}{4} = \frac{2 \times 3}{3 \times 4} = \frac{6}{12}$, now, simplify: $\frac{6}{12} = \frac{6 \div 6}{12 \div 6} = \frac{1}{2}$)

6) $\frac{4}{7}$ (Keep the first fraction, change the division sign to multiplication, and flip the numerator and denominator of the second fraction.

Then: $\frac{3}{7} \div \frac{3}{4} = \frac{3}{7} \times \frac{4}{3} = \frac{3 \times 4}{7 \times 3} = \frac{12}{21}$, now, simplify: $\frac{12}{21} = \frac{12 \div 3}{21 \div 3} = \frac{4}{7}$)

Adding Mixed Numbers

1) $8\frac{1}{6}$ (Rewriting our equation with parts separated, $5 + \frac{2}{3} + 2 + \frac{1}{2}$. Add the whole number parts: $5 + 2 = 7$. Add the fraction parts: $\frac{2}{3} + \frac{1}{2} = \frac{4}{6} + \frac{3}{6} = \frac{7}{6}$. Convert the improper fraction into a mixed number: $\frac{7}{6} = 1\frac{1}{6}$. Now, combine the whole and fraction parts: $7 + 1\frac{1}{6} = 8\frac{1}{6}$)

2) 8 (Rewriting our equation with parts separated, $3 + \frac{1}{2} + 4 + \frac{1}{2}$. Add the whole number parts: $3 + 4 = 7$. Add the fraction parts: $\frac{1}{2} + \frac{1}{2} = \frac{1+1}{2} = \frac{2}{2} = 1$. Now, combine the whole parts: $7 + 1 = 8$)

3) $4\frac{1}{2}$ (Rewriting our equation with parts separated, $2 + \frac{3}{8} + 2 + \frac{1}{8}$. Add the whole number parts: $2 + 2 = 4$. Add the fraction parts: $\frac{3}{8} + \frac{1}{8} = \frac{3}{8} + \frac{1}{8} = \frac{4}{8}$. simplify fraction: $\frac{4 \div 4}{8 \div 4} = \frac{1}{2}$. Now, combine the whole and fraction parts:

$4 + \frac{1}{2} = 4\frac{1}{2}$)

4) $10\frac{3}{14}$ (Rewriting our equation with parts separated, $4+\frac{1}{7}+6+\frac{1}{14}$. Add the whole number parts: $4+6=10$. Add the fraction parts: $\frac{1}{7}+\frac{1}{14}=\frac{2}{14}+\frac{1}{14}=\frac{3}{14}$. Now, combine the whole and fraction parts: $10+\frac{3}{14}=10\frac{3}{14}$)

5) $8\frac{7}{15}$ (Rewriting our equation with parts separated, $7+\frac{1}{5}+1+\frac{4}{15}$. Add the whole number parts: $7+1=8$. Add the fraction parts: $\frac{1}{5}+\frac{4}{15}=\frac{3}{15}+\frac{4}{15}=\frac{7}{15}$. Now, combine the whole and fraction parts: $8+\frac{7}{15}=8\frac{7}{15}$)

6) $7\frac{1}{12}$ (Rewriting our equation with parts separated, $3+\frac{1}{3}+3+\frac{3}{4}$. Add the whole number parts: $3+3=6$. Add the fraction parts: $\frac{1}{3}+\frac{3}{4}=\frac{4}{12}+\frac{9}{12}=\frac{13}{12}$. Convert the improper fraction into a mixed number: $\frac{13}{12}=1\frac{1}{12}$. Now, combine the whole and fraction parts: $6+1\frac{1}{12}=7\frac{1}{12}$)

Subtracting Mixed Numbers

1) $\frac{4}{3}$ (Convert mixed numbers into fractions: $2\frac{2}{3}=\frac{2\times3+2}{3}=\frac{8}{3}$ and $1\frac{1}{3}=\frac{1\times3+1}{3}=\frac{4}{3}$, Then Subtract the two fractions: $\frac{8}{3}-\frac{4}{3}=\frac{8-4}{3}=\frac{4}{3}$)

2) 3 (Convert mixed numbers into fractions: $6\frac{1}{2}=\frac{6\times2+1}{2}=\frac{13}{2}$ and $3\frac{1}{2}=\frac{3\times2+1}{2}=\frac{7}{2}$, Then Subtract the two fractions: $\frac{13}{2}-\frac{7}{2}=\frac{13-7}{2}=\frac{6}{2}=3$)

3) $\frac{1}{8}$ (Convert mixed numbers into fractions: $2\frac{3}{8}=\frac{2\times8+3}{8}=\frac{19}{8}$ and $2\frac{1}{4}=\frac{2\times4+1}{4}=\frac{9}{4}$. These two fractions are "unlike" fractions. (they have different denominators). Find equivalent fractions with the same denominator. Use this formula: $\frac{a}{b}-\frac{c}{d}=\frac{ad-cb}{bd}=\frac{19}{8}-\frac{9}{4}=\frac{19-(9)(2)}{8}=\frac{19-18}{8}=\frac{1}{8}$)

4) $1\frac{1}{2}$ (Convert mixed numbers into fractions: $5\frac{1}{3}=\frac{5\times3+1}{3}=\frac{16}{3}$ and $3\frac{5}{6}=\frac{3\times6+5}{6}=\frac{23}{6}$. These two fractions are "unlike" fractions. (they have different denominators). Find equivalent fractions with the same denominator. Use this formula: $\frac{a}{b}-\frac{c}{d}=\frac{ad-cb}{bd}=\frac{16}{3}-\frac{23}{6}=\frac{(16)(2)-23}{6}=\frac{32-23}{6}=\frac{9}{6}$.simplify fraction: $\frac{9\div3}{6\div3}=\frac{3}{2}$. Convert the improper fraction into a mixed number: $\frac{3}{2}=1\frac{1}{2}$.)

5) $3\frac{1}{5}$ (Convert mixed numbers into fractions: $5\frac{1}{2}=\frac{5\times2+1}{2}=\frac{11}{2}$ and $2\frac{3}{10}=\frac{2\times10+3}{10}=\frac{23}{10}$.These two fractions are "unlike" fractions. Find equivalent fractions with the same denominator: $\frac{11}{2}-\frac{23}{10}=\frac{(11)(5)-23}{10}=\frac{55-23}{10}=\frac{32}{10}$.simplify fraction: $\frac{32\div2}{10\div2}=\frac{16}{5}$. Convert the improper fraction into a mixed number: $\frac{16}{5}=3\frac{1}{5}$)

6) $6\frac{5}{12}$ (Convert mixed numbers into fractions: $10\frac{2}{3}=\frac{10\times3+2}{3}=\frac{32}{3}$ and $4\frac{1}{4}=\frac{4\times4+1}{4}=\frac{17}{4}$. Then: $\frac{32}{3}-\frac{17}{4}=\frac{(32)(4)-(17)(3)}{3\times4}=\frac{128-51}{12}=\frac{77}{12}=6\frac{5}{12}$)

Multiplying Mixed Numbers

1) $8\frac{4}{5}$ (Convert mixed numbers into fractions, $3\frac{2}{3}=\frac{3\times3+2}{3}=\frac{11}{3}$ and $2\frac{2}{5}=\frac{2\times5+2}{5}=\frac{12}{5}$. Apply the fractions rule for multiplication: $\frac{11}{3}\times\frac{12}{5}=\frac{11\times12}{3\times5}=\frac{132}{15}=\frac{44}{5}$. The answer is an improper fraction. Convert it into a mixed number. $\frac{44}{5}=8\frac{4}{5}$)

2) $18\frac{1}{3}$ (Converting mixed numbers into fractions, $5\frac{1}{2}\times3\frac{1}{3}=\frac{11}{2}\times\frac{10}{3}$ .Apply the fractions rule for multiplication: $\frac{11}{2}\times\frac{10}{3}=\frac{11\times10}{2\times3}=\frac{110}{6}=\frac{55}{3}=18\frac{1}{3}$)

3) $6\frac{4}{5}$ (Converting mixed numbers into fractions, $4\frac{1}{4}\times 1\frac{3}{5}=\frac{17}{4}\times\frac{8}{5}$. Apply the fractions rule for multiplication: $\frac{17}{4}\times\frac{8}{5}=\frac{17\times 8}{4\times 5}=\frac{136}{20}=\frac{34}{5}=6\frac{4}{5}$)

4) $22\frac{1}{2}$ (Converting mixed numbers into fractions, $10\frac{1}{8}\times 2\frac{2}{9}=\frac{81}{8}\times\frac{20}{9}$. Apply the fractions rule for multiplication: $\frac{81}{8}\times\frac{20}{9}=\frac{81\times 20}{8\times 9}=\frac{1{,}620}{72}=\frac{45}{2}=22\frac{1}{2}$)

5) $19\frac{4}{5}$ (Converting mixed numbers into fractions, $7\frac{1}{5}\times 2\frac{3}{4}=\frac{36}{5}\times\frac{11}{4}$. Apply the fractions rule for multiplication: $\frac{36}{5}\times\frac{11}{4}=\frac{36\times 11}{5\times 4}=\frac{396}{20}=\frac{99}{5}=19\frac{4}{5}$)

6) 8 (Converting mixed numbers into fractions, $7\frac{1}{3}\times 1\frac{1}{11}=\frac{22}{3}\times\frac{12}{11}$. Apply the fractions rule for multiplication: $\frac{22}{3}\times\frac{12}{11}=\frac{22\times 12}{3\times 11}=\frac{264}{33}=8$)

Dividing Mixed Numbers

1) $\frac{8}{11}$ (Convert mixed numbers into fractions: $2\frac{2}{3}=\frac{2\times 3+2}{3}=\frac{8}{3}$ and $3\frac{2}{3}=\frac{3\times 3+2}{3}=\frac{11}{3}$. Keep, Change, Flip: $\frac{8}{3}\div\frac{11}{3}=\frac{8}{3}\times\frac{3}{11}=\frac{8\times 3}{3\times 11}=\frac{24\div 3}{33\div 3}=\frac{8}{11}$)

2) $6\frac{5}{6}$ (Convert mixed numbers to fractions, then solve: $10\frac{1}{4}\div 1\frac{1}{2}=\frac{41}{4}\div\frac{3}{2}=\frac{41}{4}\times\frac{2}{3}=\frac{82}{12}=\frac{41}{6}=6\frac{5}{6}$)

3) $\frac{5}{8}$ (Convert mixed numbers to fractions, then solve: $1\frac{3}{7}\div 2\frac{2}{7}=\frac{10}{7}\div\frac{16}{7}=\frac{10}{7}\times\frac{7}{16}=\frac{70}{112}=\frac{5}{8}$)

4) $1\frac{1}{4}$ (Convert mixed numbers to fractions, then solve: $4\frac{1}{6}\div 3\frac{1}{3}=\frac{25}{6}\div\frac{10}{3}=\frac{25}{6}\times\frac{3}{10}=\frac{75}{60}=\frac{5}{4}=1\frac{1}{4}$)

5) $2\frac{10}{21}$ (Convert mixed numbers to fractions, then solve: $5\frac{1}{5}\div 2\frac{1}{10}=\frac{26}{5}\div\frac{21}{10}=\frac{26}{5}\times\frac{10}{21}=\frac{260}{105}=\frac{52}{21}=2\frac{10}{21}$)

6) $\frac{1}{2}$ (Convert mixed numbers to fractions, then solve: $1\frac{3}{8}\div 2\frac{3}{4}=\frac{11}{8}\div\frac{11}{4}=\frac{11}{8}\times\frac{4}{11}=\frac{1}{2}$)

CHAPTER 2 Decimals

Math topics that you'll learn in this chapter:

- ☑ Comparing Decimals
- ☑ Rounding Decimals
- ☑ Adding and Subtracting Decimals
- ☑ Multiplying and Dividing Decimals

Comparing Decimals

- A decimal is a fraction written in a special form. For example, instead of writing $\frac{1}{2}$, you can write: 0.5
- A Decimal Number contains a Decimal Point. It separates the whole number part from the fractional part of a decimal number.
- Let's review decimal place values: Example: 45.3861

  4: tens 5: ones 3: tenths

  8: hundredths 6: thousandths 1: tens thousandths
- To compare two decimals, compare each digit of two decimals in the same place value. Start from the left. Compare hundreds, tens, ones, tenth, hundredth, etc.
- To compare numbers, use these symbols:

  Equal to $=$ Less than $<$ Greater than $>$

  Greater than or equal $\geq$ Less than or equal $\leq$

Examples:

Example 1. Compare 0.03 and 0.30.

Solution: 0.30 is greater than 0.03, because the tenth place of 0.30 is 3, but the tenth place of 0.03 is zero. Then: $0.03 < 0.30$

Example 2. Compare 0.0917 and 0.217.

Solution: 0.217 is greater than 0.0917, because the tenth place of 0.217 is 2, but the tenth place of 0.0917 is zero. Then: $0.0917 < 0.217$

Practices:

Write the correct comparison symbol (>, < or =).

1) 0.90 ☐ 0.090
2) 0.067 ☐ 0.67
3) 5.40 ☐ 5.5
4) 7.09 ☐ 7.090
5) 0.431 ☐ 0.352
6) 8.08 ☐ 0.88
7) 4.04 ☐ 4.4
8) 2.04 ☐ 2.3

Rounding Decimals

- We can round decimals to a certain accuracy or number of decimal places. This is used to make calculations easier to do and results easier to understand when exact values are not too important.
- First, you'll need to remember your place values: For example: 12.4869

  1: tens 2: ones 4: tenths

  8: hundredths 6: thousandths 9: tens thousandths
- To round a decimal, first find the place value you'll round to.
- Find the digit to the right of the place value you're rounding to. If it is 5 or bigger, add 1 to the place value you're rounding to and remove all digits on its right side. If the digit to the right of the place value is less than 5, keep the place value and remove all digits on the right.

Examples:

Example 1. Round 4.3679 to the thousandths place value.
Solution: First, look at the next place value to the right, (tens thousandths). It's 9 and it is greater than 5. Thus add 1 to the digit in the thousandth place. The thousandth place is 7. $\rightarrow 7 + 1 = 8$, then, the answer is 4.368

Example 2. Round 1.5237 to the nearest hundredth.
Solution: First, look at the digit to the right of the hundredth (thousandths place value). It's 3 and it is less than 5, thus remove all the digits to the right of hundredth place. Then, the answer is 1.52

Practices:

Round each decimal to the nearest whole number.

1) 25.93
2) 16.3
3) 6.77
4) 2.5
5) 27.95
6) 16.44

Round each decimal to the nearest tenth.

7) 26.826
8) 33.729
9) 17.779
10) 53.66
11) 21.259
12) 88.416

Adding and Subtracting Decimals

- Line up the decimal numbers.
- Add zeros to have the same number of digits for both numbers if necessary.
- Remember your place values: For example: 73.5196

  7: tens 3: ones 5: tenths

  1: hundredths 9: thousandths 6: tens thousandths

- Add or subtract using column addition or subtraction.

Examples:

Example 1. Add. $1.7 + 4.12$

Solution: First, line up the numbers: $\frac{\begin{array}{r}1.7\\+4.12\end{array}}{}$ → Add a zero to have the same number of digits for both numbers. $\frac{\begin{array}{r}1.70\\+4.12\end{array}}{}$ → Start with the hundredths place: $0 + 2 = 2$, $\frac{\begin{array}{r}1.70\\+4.12\end{array}}{2}$ → Continue with tenths place: $7 + 1 = 8$, $\frac{\begin{array}{r}1.70\\+4.12\end{array}}{.82}$ → Add the ones place: $4 + 1 = 5$, $\frac{\begin{array}{r}1.70\\+4.12\end{array}}{5.82}$ → The answer is 5.82.

Example 2. Find the difference. $5.58 - 4.23$

Solution: First, line up the numbers: $\frac{\begin{array}{r}5.58\\-4.23\end{array}}{}$ → Start with the hundredths place: $8 - 3 = 5$, $\frac{\begin{array}{r}5.58\\-4.23\end{array}}{5}$ → Continue with tenths place. $5 - 2 = 3$, $\frac{\begin{array}{r}5.58\\-4.23\end{array}}{.35}$ → Subtract the ones place. $5 - 4 = 1$, $\frac{\begin{array}{r}5.58\\-4.23\end{array}}{1.35}$

Practices:

Find the sum or difference.

1) $23.66 - 13.25 =$

2) $12.48 + 27.11 =$

3) $74.31 + 13.47 =$

4) $67.94 - 36.43 =$

5) $87.88 - 43.22 =$

6) $57.41 + 41.37 =$

Multiplying and Dividing Decimals

For multiplying decimals:

- Ignore the decimal point and set up and multiply the numbers as you do with whole numbers.
- Count the total number of decimal places in both of the factors.
- Place the decimal point in the product.

For dividing decimals:

- If the divisor is not a whole number, move the decimal point to the right to make it a whole number. Do the same for the dividend.
- Divide similar to whole numbers.

Examples:

Example 1. Find the product. $0.65 \times 0.24 =$

Solution: Set up and multiply the numbers as you do with whole numbers. Line up the numbers: $\frac{65}{\times 24}$ → Start with the ones place then continue with other digits → $\begin{array}{r} 65 \\ \times 24 \\ \hline 1{,}560 \end{array}$. Count the total number of decimal places in both of the factors. There are four decimal's digits. (two for each factor 0.65 and 0.24) Then: $0.65 \times 0.24 = 0.1560 = 0.156$

Example 2. Find the quotient. $1.20 \div 0.4 =$

Solution: The divisor is not a whole number. Multiply it by 10 to get 4: → $0.4 \times 10 = 4$

Do the same for the dividend to get 12. → $1.20 \times 10 = 12$

Now, divide $12 \div 4 = 3$. The answer is 3.

Practices:

Find the product and quotient.

1) $0.3 \times 0.8 =$	4) $0.88 \times 0.7 =$	7) $6.3 \div 0.9 =$
2) $3.7 \times 0.4 =$	5) $2.75 \div 0.25 =$	8) $30.1 \div 0.07 =$
3) $2.75 \times 0.3 =$	6) $5.7 \div 0.3 =$	

Chapter 2: Answers

Comparing Decimals

1) $>$ (0.90 is greater than 0.090, because the tenth place of 0.90 is 9, but the tenth place of 0.090 is zero. Then: $0.90 > 0.090$)
2) $<$ (0.67 is greater than 0.067, because the tenth place of 0.67 is 6, but the tenth place of 0.067 is zero. Then: $0.067 < 0.67$)
3) $<$ (5.5 is greater than 5.40, because the tenth place of 5.5 is 5, but the tenth place of 5.40 is 4. Then: $5.40 < 5.5$)
4) $=$ (Two numbers are equal)
5) $>$(0.431 is greater than 0.352, because the tenth place of 0.431 is 4, but the tenth place of 0.352 is 3. Then: $0.431 > 0.352$)
6) $>$ (8.08 is greater than 0.88, because the ones place of 8.08 is 8, but the ones place of 0.88 is zero. Then: $8.08 > 0.88$)
7) $<$ (4.4 is greater than 4.04, because the tenth place of 4.4 is 4, but the tenth place of 4.04 is 0. Then: $4.04 < 4.4$)
8) $<$ (2.3 is greater than 2.04, because the tenth place of 2.3 is 3, but the tenth place of 2.04 is zero. Then: $2.04 < 2.3$)

Rounding Decimals

1) 26 (Look at tenths place. It's 9 and it is greater than 5. Thus add 1 to the digit in the ones place. $5 + 1 = 6$, then, the answer is 26)
2) 16 (Tenths place is 3 and it is less than 5. Thus, remove all the digits to the right. Then, the answer is 16)
3) 7 (Tenths place is 7 and it is greater than 5. Thus add 1 to the digit in the ones place. $6 + 1 = 7$, then, the answer is 7)

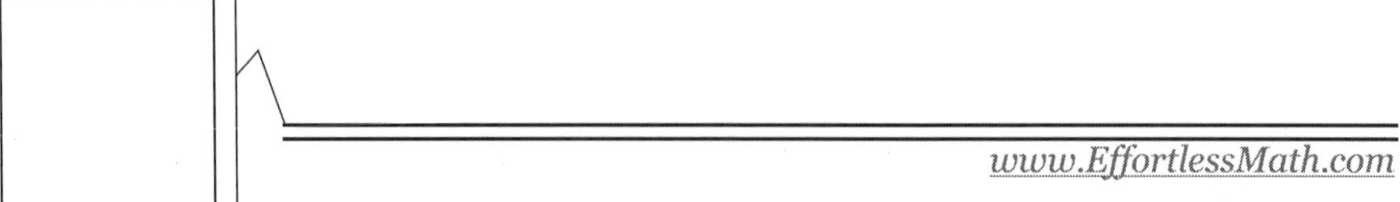

4) 3 (Tenths place is 5. Thus add 1 to the digit in the ones place. $2+1=4$, then, the answer is 3)

5) 28 (Tenths place is 9 and it is greater than 5. Thus add 1 to the digit in the ones place. $27+1=28$, then, the answer is 28)

6) 16 (Tenths place is 4 and it is less than 5. Thus, remove all the digits to the right. Then, the answer is 16)

7) 26.8 (Look at the next place value to the right, (hundredths). It's 2 and it is less than 5. Thus, remove all the digits to the right. Then, the answer is 26.8)

8) 33.7 (Hundredths place is 2 and it's less than 5. Thus, remove all the digits to the right. Then, the answer is 33.7)

9) 17.8 (Hundredths place is 7 and it's greater than 5. Thus add 1 to the digit in the tenths place. $7+1=8$, then, the answer is 17.8)

10) 53.7 (Hundredths place is 6 and it's greater than 5. Thus add 1 to the digit in the tenths place. $6+1=7$, then, the answer is 53.7)

11) 21.3 (Hundredths place is 5. Thus add 1 to the digit in the tenths place. $2+1=3$, then, the answer is 21.3)

12) 88.4 (Hundredths place is 1 and it's less than 5. Thus, remove all the digits to the right. Then, the answer is 88.4)

Adding and Subtracting Decimals

1) 10.41 ($\begin{array}{r}23.66\\ \underline{-13.25}\\ \end{array} \rightarrow 6-5=1 \rightarrow \begin{array}{r}23.66\\ \underline{-13.25}\\ 1\end{array} \rightarrow 6-2=4 \rightarrow \begin{array}{r}23.66\\ \underline{-13.25}\\ 41\end{array} \rightarrow 3-3=0 \rightarrow$

$\begin{array}{r}23.66\\ \underline{-13.25}\\ 0.41\end{array} \rightarrow 2-1=1 \rightarrow \begin{array}{r}23.66\\ \underline{-13.25}\\ 10.41\end{array}$)

2) 39.59 ($\begin{array}{r}12.48\\ \underline{+27.11}\\ \end{array} \rightarrow 8+1=9 \rightarrow \begin{array}{r}12.48\\ \underline{+27.11}\\ 9\end{array} \rightarrow 4+1=5 \rightarrow \begin{array}{r}12.48\\ \underline{+27.11}\\ 59\end{array} \rightarrow 2+7=9 \rightarrow$

$\begin{array}{r}12.48\\ \underline{+27.11}\\ 9.59\end{array} \rightarrow 2+1=3 \rightarrow \begin{array}{r}12.48\\ \underline{+27.11}\\ 39.59\end{array}$)

3) 87.78 ($\begin{array}{r}74.31\\+13.47\\\hline\end{array} \rightarrow 1+7=8 \rightarrow \begin{array}{r}74.31\\+13.47\\\hline 8\end{array} \rightarrow 3+4=7 \rightarrow \begin{array}{r}74.31\\+13.47\\\hline 78\end{array} \rightarrow 4+3=7 \rightarrow$

$\begin{array}{r}74.31\\+13.47\\\hline 7.78\end{array} \rightarrow 7+1=8 \rightarrow \begin{array}{r}74.31\\+13.47\\\hline 87.78\end{array}$)

4) 31.51 ($\begin{array}{r}67.94\\-36.43\\\hline\end{array} \rightarrow 4-3=1 \rightarrow \begin{array}{r}67.94\\-36.43\\\hline 1\end{array} \rightarrow 9-4=5 \rightarrow \begin{array}{r}67.94\\-36.43\\\hline 51\end{array} \rightarrow 7-6=1 \rightarrow$

$\begin{array}{r}67.94\\-36.43\\\hline 1.51\end{array} \rightarrow 6-3=3 \rightarrow \begin{array}{r}67.94\\-36.43\\\hline 31.51\end{array}$)

5) 44.66 ($\begin{array}{r}87.88\\-43.22\\\hline\end{array} \rightarrow 8-2=6 \rightarrow \begin{array}{r}87.88\\-43.22\\\hline 6\end{array} \rightarrow 8-2=6 \rightarrow \begin{array}{r}87.88\\-43.22\\\hline 66\end{array} \rightarrow 7-3=4 \rightarrow$

$\begin{array}{r}87.88\\-43.22\\\hline 4.66\end{array} \rightarrow 8-4=4 \rightarrow \begin{array}{r}87.88\\-43.22\\\hline 44.66\end{array}$)

6) 98.78 ($\begin{array}{r}57.41\\+41.37\\\hline\end{array} \rightarrow 1+7=8 \rightarrow \begin{array}{r}57.41\\+41.37\\\hline 8\end{array} \rightarrow 4+3=7 \rightarrow \begin{array}{r}57.41\\+41.37\\\hline 78\end{array} \rightarrow 7+1=8 \rightarrow$

$\begin{array}{r}57.41\\+41.37\\\hline 8.78\end{array} \rightarrow 5+4=9 \rightarrow \begin{array}{r}57.41\\+41.37\\\hline 98.78\end{array}$

Multiplying and Dividing Decimals

1) 0.24 (Line up the numbers: $\begin{array}{r}8\\\times 3\\\hline 24\end{array}$ →Count the total number of decimal places in both of the factors. There are two decimals digits. (one for each factor 0.8 and 0.3) Then: $0.8 \times 0.3 = 0.24$)

2) 1.48 ($\begin{array}{r}37\\\times 4\\\hline 148\end{array}$ →Count the total number of decimal places in both of the factors. There are two decimals digits. (one for each factor 3.7 and 0.4) Then: $3.7 \times 0.4 = 1.48$)

3) 0.825 ($\begin{array}{r}275\\\times\ \ 3\\\hline 825\end{array}$ → There are three decimals digits. Then: $2.75 \times 0.3 = 0.825$)

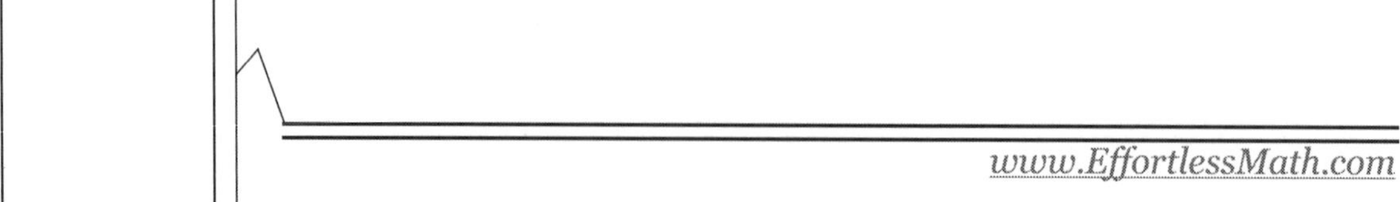

4) 0.616 ($\begin{array}{r} 88 \\ \underline{\times\ 7} \\ 616 \end{array}$ → There are three decimals digits. Then: $0.88 \times 0.7 = 0.616$)

5) 11 (The divisor is not a whole number. Multiply it by 100 to get 25. Do the same for the dividend to get 275. Now, divide: $275 \div 25 = 11$)

6) 19 (The divisor is not a whole number. Multiply it by 10 to get 3. Do the same for the dividend to get 57. Now, divide: $57 \div 3 = 19$)

7) 7 (The divisor is not a whole number. Multiply it by 10 to get 9. Do the same for the dividend to get 63. Now, divide: $63 \div 9 = 7$)

8) 430 (The divisor is not a whole number. Multiply it by 100 to get 7. Do the same for the dividend to get $3{,}010$. Now, divide: $3{,}010 \div 7 = 430$)

CHAPTER

3 Integers and Order of Operations

Math topics that you'll learn in this chapter:

- ☑ Adding and Subtracting Integers
- ☑ Multiplying and Dividing Integers
- ☑ Order of Operations
- ☑ Integers and Absolute Value

Adding and Subtracting Integers

- Integers include zero, counting numbers, and the negative of the counting numbers. $\{\ldots, -3, -2, -1, 0, 1, 2, 3, \ldots\}$
- Add a positive integer by moving to the right on the number line. (you will get a bigger number)
- Add a negative integer by moving to the left on the number line. (you will get a smaller number)
- Subtract an integer by adding its opposite.

Number line

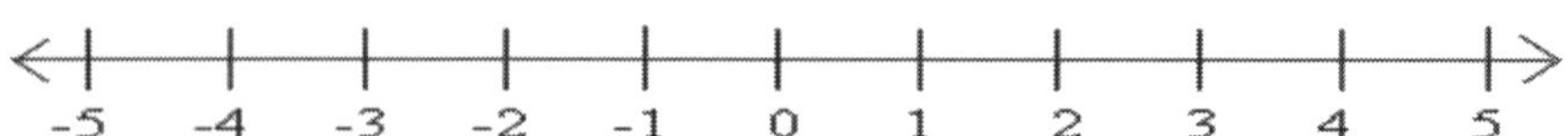

Examples:

Example 1. Solve. $(-2) - (-8) =$

Solution: Keep the first number and convert the sign of the second number to its opposite (change subtraction into addition). Then: $(-2) + 8 = 6$

Example 2. Solve. $4 + (5 - 10) =$

Solution: First, subtract the numbers in brackets, $5 - 10 = -5$.
Then: $4 + (-5) = \rightarrow$ change addition into subtraction: $4 - 5 = -1$

Practices:

Find each sum or difference.

1) $18 + (-7) =$
2) $(-21) + (-13) =$
3) $(-11) - (-26) =$
4) $67 + (-23) =$
5) $(-9) + (-15) + 6 =$
6) $59 + (-22) + 14 =$
7) $(-14) - (-6) =$
8) $15 - (-33) =$

Multiplying and Dividing Integers

Use the following rules for multiplying and dividing integers:

- (negative) × (negative) = positive
- (negative) ÷ (negative) = positive
- (negative) × (positive) = negative
- (negative) ÷ (positive) = negative
- (positive) × (positive) = positive
- (positive) ÷ (negative) = negative

Examples:

Example 1. Solve. $3 \times (-4) =$

Solution: Use this rule: (positive) × (negative) = negative

Then: $(3) \times (-4) = -12$

Example 2. Solve. $(-3) + (-24 \div 3) =$

Solution: First, divide -24 by 3, the numbers in brackets, use this rule:

(negative) ÷ (positive) = negative. Then: $-24 \div 3 = -8$

$(-3) + (-24 \div 3) = (-3) + (-8) = -3 - 8 = -11$

Practices:

Find each product or quotient.

1) $(-6) \times (-9) =$
2) $(-5) \times (-20) =$
3) $-(4) \times (-7) \times 5 =$
4) $(16 - 3) \times (-9) =$
5) $32 \div (-8) =$
6) $(-66) \div 11 =$
7) $(-36) \div (-9) =$
8) $(-49) \div (-7) =$

Find more at bit.ly/3pjQW98

Order of Operations

- In Mathematics, "operations" are addition, subtraction, multiplication, division, exponentiation (written as b^n), and grouping.
- When there is more than one math operation in an expression, use PEMDAS: (to memorize this rule, remember the phrase "Please Excuse My Dear Aunt Sally".)
  - Parentheses
  - Exponents
  - Multiplication and Division (from left to right)
  - Addition and Subtraction (from left to right)

Examples:

Example 1. Calculate. $(2 + 6) \div (2^2 \div 4) =$

Solution: First, simplify inside parentheses:
$(8) \div (4 \div 4) = (8) \div (1)$, Then: $(8) \div (1) = 8$

Example 2. Solve. $(6 \times 5) - (14 - 5) =$

Solution: First, calculate within parentheses: $(6 \times 5) - (14 - 5) = (30) - (9)$, Then: $(30) - (9) = 21$

Practices:

Evaluate each expression

1) $6 + (3 \times 4) =$
2) $10 - (2 \times 3) =$
3) $(18 \times 2) + 15 =$
4) $(17 - 2) - (3 \times 3) =$
5) $30 + (18 \div 6) =$
6) $(12 \times 10) \div 4 =$
7) $(24 \div 4) \times (-5) =$
8) $(7 \times 8) + (24 - 12) =$

Integers and Absolute Value

- The absolute value of a number is its distance from zero, in either direction, on the number line. For example, the distance of 9 and -9 from zero on the number line is 9.
- The absolute value of an integer is the numerical value without its sign. (negative or positive)
- The vertical bar is used for absolute value as in $|x|$.
- The absolute value of a number is never negative; because it only shows, "how far the number is from zero".

Examples:

Example 1. Calculate. $|14-2|\times 5=$

Solution: First, solve $|14-2|$, $\rightarrow |14-2|=|12|$, the absolute value of 12 is 12, $|12|=12$, Then: $12\times 5=60$

Example 2. Solve. $\frac{|-24|}{4}\times|5-7|=$

Solution: First, find $|-24|$ $\rightarrow$ the absolute value of -24 is 24. Then: $|-24|=24$, $\frac{24}{4}\times|5-7|=$
Now, calculate $|5-7|$, $\rightarrow$ $|5-7|=|-2|$, the absolute value of -2 is 2. $|-2|=2$
Then: $\frac{24}{4}\times 2=6\times 2=12$

Practices:

Evaluate the value.

1) $12-|4-7|-|-1|=$
2) $|-12|-\frac{|-4|}{2}=$
3) $\frac{|-18|}{3}\times|-5|=$
4) $\frac{|10\times-3|}{3}\times\frac{|-16|}{4}=$
5) $|11\times-2|+\frac{|-15|}{3}=$
6) $\frac{|-33|}{3}\times\frac{|-16|}{8}=$
7) $|-10+4|\times\frac{|-3\times4|}{12}=$
8) $\frac{|-7\times6|}{3}\times|-6|=$

Find more at bit.ly/3aD521u

Chapter 3: Answers

Adding and Subtracting Integers

1) 11 (Change addition into subtraction: $18 - 7 = 11$)
2) -34 (Keep the first number and change addition into subtraction: $-21 - 13 = -34$)
3) 15 (Keep the first number and convert the sign of the second number to its opposite. change subtraction into addition. Then: $(-11) + 26 = 15$)
4) 44 (Change addition into subtraction: $67 - 23 = 44$)
5) -18 (Keep the first number and change the first addition into subtraction. Then: $(-9) - 15 + 6 = -18$)
6) 51 (Keep the first number and change the first addition into subtraction. Then: $59 - 22 + 14 = 51$)
7) -8 (Keep the first number and convert the sign of the second number to its opposite. change subtraction into addition. Then: $(-14) + 6 = -8$)
8) 48 (Keep the first number and convert the sign of the second number to its opposite. change subtraction into addition. Then: $15 + 33 = 48$)

Multiplying and Dividing Integers

1) 54 (Use this rule: (negative) × (negative) = positive. Then: $(-6) \times (-9) = 54$)
2) 100 (Use this rule: (negative) × (negative) = positive. Then: $(-5) \times (-20) = 100$)
3) 140 (Use this rule for the first two numbers: (negative) × (negative) = positive. Then: $(-4) \times (-7) = 28$. Using this rule, we multiply the number obtained in the third decimal: (positive) × (positive) = positive. Then: $28 \times 5 = 140$)

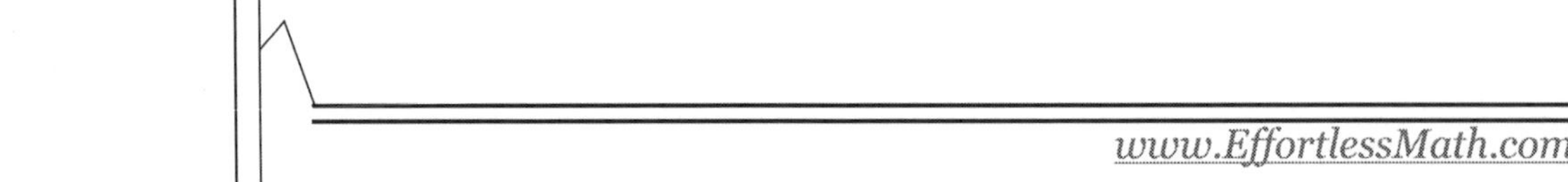

4) -117 (Subtract the numbers in brackets, $16 - 3 = 13 \rightarrow 13 \times (-9) =$)

Now use this rule: (positive) × (negative) = negative → $13 \times (-9) = -117$)

5) -4 (Use this rule: (positive) ÷ (negative) = negative. Then: $32 \div (-8) = -4$)

6) -6 (Use this rule: (negative) ÷ (positive) = negative. Then: $(-66) \div 11 = -6$)

7) 4 (Use this rule: (negative) ÷ (negative) = positive. Then: $(-36) \div (-9) = 4$)

8) 7 (Use this rule: (negative) ÷ (negative) = positive. Then: $(-49) \div (-7) = 7$)

Order of Operations

1) 18 (First, simplify inside parentheses: $6 + (3 \times 4) = 6 + (12)$, Then: $6 + 12 = 18$)

2) 4 (First, simplify inside parentheses: $10 - (2 \times 3) = 10 - (6)$, Then: $10 - 6 = 4$)

3) 51 (First, simplify inside parentheses: $(18 \times 2) + 15 = (36) + 15$, Then: $36 + 15 = 51$)

4) 6 (First, simplify inside parentheses: $(17 - 2) - (3 \times 3) = (15) - (9)$, Then: $15 - 9 = 6$)

5) 33 (First, simplify inside parentheses: $30 + (18 \div 6) = 30 + (3)$, Then: $30 + 3 = 33$)

6) 30 (First, simplify inside parentheses: $(12 \times 10) \div 4 = (120) \div 4$, Then: $120 \div 4 = 30$)

7) -30(First, simplify inside parentheses: $(24 \div 4) \times (-5) = (6) \times (-5)$, Then: $6 \times -5 = -30$)

8) 68 (First, simplify inside parentheses: $(7 \times 8) + (24 - 12) = (56) + (12)$, Then: $56 + 12 = 68$)

Integers and Absolute Value

1) 8 (First, find $|4-7|=|-3|$ → the absolute value of -3 is 3. Now, the absolute value of -1 is 1. Then: $12-3-1=8$)
2) 10 (First, find $|-12|$ → the absolute value of -12 is 12. Now, calculate $|-4|$ → the absolute value of -4 is $4 \rightarrow \frac{4}{2}=2$.Then: $12-2=10$)
3) 30 (First, find $|-18|$ → the absolute value of -18 is $18 \rightarrow \frac{18}{3}=6$. Now, calculate $|-5|$ → the absolute value of -5 is 5.Then: $6\times 5=30$)
4) 40 (First, find $|10\times -3| \rightarrow |-30|$ the absolute value of -30 is $30 \rightarrow \frac{30}{3}=10$. Now, calculate $|-16|$ → the absolute value of -16 is $16 \rightarrow \frac{16}{4}=4$.Then: $10\times 4=40$)
5) 27 (First, find $|11\times -2| \rightarrow |-22|$ the absolute value of -22 is 22. Now, calculate $|-15|$ → the absolute value of -15 is $15 \rightarrow \frac{15}{3}=5$ .Then: $22+5=27$)
6) 22 (First, find $|-33|$ the absolute value of -33 is $33 \rightarrow \frac{33}{3}=11$. Now, calculate $|-16|$ → the absolute value of -16 is $16 \rightarrow \frac{16}{8}=2$ .Then: $11\times 2=22$)
7) 6 (First, find $|-10+4| \rightarrow |-6|$ the absolute value of -6 is 6. Now, calculate $|-3\times 4| \rightarrow |-12|$ the absolute value of -12 is $12 \rightarrow \frac{12}{12}=1$.Then: $6\times 1=6$)
8) 84 (First, find $|-7\times 6| \rightarrow |-42|$ the absolute value of -42 is $42 \rightarrow \frac{42}{3}=14$. Now, calculate $|-6|$ the absolute value of -6 is 6.Then: $14\times 6=84$)

CHAPTER

4 Ratios and Proportions

Math topics that you'll learn in this chapter:

- ☑ Simplifying Ratios
- ☑ Proportional Ratios
- ☑ Similarity and Ratios

Simplifying Ratios

- Ratios are used to make comparisons between two numbers.
- Ratios can be written as a fraction, using the word "to", or with a colon. Example: $\frac{3}{4}$ or "3 to 4" or $3:4$
- You can calculate equivalent ratios by multiplying or dividing both sides of the ratio by the same number.

Examples:

Example 1. Simplify. $8:2 =$

Solution: Both numbers 8 and 2 are divisible by $2 \Rightarrow 8 \div 2 = 4$, $2 \div 2 = 1$, Then: $8:2 = 4:1$

Example 2. Simplify. $\frac{9}{33} =$

Solution: Both numbers 9 and 33 are divisible by $3 \Rightarrow 33 \div 3 = 11$, $9 \div 3 = 3$, Then: $\frac{9}{33} = \frac{3}{11}$

Practices:

Reduce each ratio.

1) $24:4 =$ ___:___
2) $12:3 =$ ___:___
3) $4:48 =$ ___:___
4) $5:15 =$ ___:___
5) $9:120 =$ ___:___
6) $18:60 =$ ___:___
7) $16:64 =$ ___:___
8) $60:90 =$ ___:___
9) $30:80 =$ ___:___
10) $12:26 =$ ___:___
11) $63:28 =$ ___:___
12) $36:66 =$ ___:___

Proportional Ratios

- Two ratios are proportional if they represent the same relationship.
- A proportion means that two ratios are equal. It can be written in two ways: $\frac{a}{b} = \frac{c}{d}$ $\quad a:b = c:d$
- The proportion $\frac{a}{b} = \frac{c}{d}$ can be written as: $a \times d = c \times b$

Examples:

Example 1. Solve this proportion for x. $\frac{2}{5} = \frac{6}{x}$

Solution: Use cross multiplication: $\frac{2}{5} = \frac{6}{x} \Rightarrow 2 \times x = 6 \times 5 \Rightarrow 2x = 30$

Divide both sides by 2 to find x: $x = \frac{30}{2} \Rightarrow x = 15$

Example 2. If a box contains red and blue balls in the ratio of 3: 5 red to blue, how many red balls are there if 45 blue balls are in the box?

Solution: Write a proportion and solve. $\frac{3}{5} = \frac{x}{45}$

Use cross multiplication: $3 \times 45 = 5 \times x \Rightarrow 135 = 5x$

Divide to find x: $x = \frac{135}{5} \Rightarrow x = 27$. There are 27 red balls in the box.

Practices:

Solve each proportion.

1) $\frac{1}{4} = \frac{x}{44}, x =$ ____
2) $\frac{1}{7} = \frac{8}{x}, x =$ ____
3) $\frac{2}{7} = \frac{14}{x}, x =$ ____
4) $\frac{3}{10} = \frac{x}{90}, x =$ ____
5) $\frac{4}{5} = \frac{x}{65}, x =$ ____
6) $\frac{1}{6} = \frac{15}{x}, x =$ ____
7) $\frac{4}{9} = \frac{64}{x}, x =$ ____
8) $\frac{5}{12} = \frac{75}{x}, x =$ ____

Similarity and Ratios

- Two figures are similar if they have the same shape.
- Two or more figures are similar if the corresponding angles are equal, and the corresponding sides are in proportion.

Examples:

Example 1. The following triangles are similar. What is the value of the unknown side?

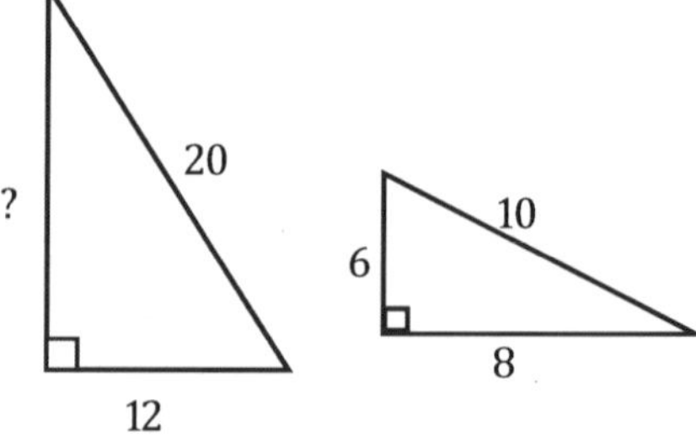

Solution: Find the corresponding sides and write a proportion.

$\frac{10}{20}=\frac{8}{x}$. Now, use the cross product to solve for x:

$\frac{10}{20}=\frac{8}{x} \rightarrow 10 \times x = 8 \times 20 \rightarrow 10x = 160$. Divide both sides by 10. Then: $10x = 160 \rightarrow x = \frac{160}{10} \rightarrow x = 16$

The missing side is 16.

Example 2. Two rectangles are similar. The first is 5 feet wide and 15 feet long. The second is 10 feet wide. What is the length of the second rectangle?

Solution: Let's put x for the length of the second rectangle. Since two rectangles are similar, their corresponding sides are in proportion. Write a proportion and solve for the missing number.

$\frac{5}{10}=\frac{15}{x} \rightarrow 5x = 10 \times 15 \rightarrow 5x = 150 \rightarrow x = \frac{150}{5} = 30$

The length of the second rectangle is 30 feet.

Practices:

Solve.

1) Two rectangles are similar. The first is 8 $feet$ wide and 28 $feet$ long. The second is 18 $feet$ wide. What is the length of the second rectangle? ________________

2) Two rectangles are similar. One is 6 $meters$ by 36 $meters$. The longer side of the second rectangle is 12 $meters$. What is the other side of the second rectangle?

____

3) A building casts a shadow 33 ft long. At the same time a girl 6 ft tall casts a shadow 3 ft long. How tall is the building? ________________

Chapter 4: Answers

Simplifying Ratios

1) $6:1$ (Both numbers 24 and 4 are divisible by $4 \Rightarrow 24 \div 4 = 6$, $4 \div 4 = 1$, Then: $24:4 = 6:1$)
2) $4:1$ (12 and 3 are divisible by $3 \Rightarrow 12 \div 3 = 4, 3 \div 3 = 1$, Then: $12:3 = 4:1$)
3) $1:12$ (4 and 48 are divisible by $4 \Rightarrow 4 \div 4 = 1, 48 \div 4 = 12$, Then: $4:48 = 1:12$)
4) $1:3$ (5 and 15 are divisible by $5 \Rightarrow 5 \div 5 = 1, 15 \div 5 = 3$, Then: $5:15 = 1:3$)
5) $3:40$ (9 and 120 are divisible by $3 \Rightarrow 9 \div 3 = 3$, $120 \div 3 = 40$, Then: $9:120 = 3:40$)
6) $3:10$ (18 and 60 are divisible by $6 \Rightarrow 18 \div 6 = 3$, $60 \div 6 = 10$, Then: $18:60 = 3:10$)
7) $1:4$ (16 and 64 are divisible by $16 \Rightarrow 16 \div 16 = 1$, $64 \div 16 = 4$, Then: $16:64 = 1:4$)
8) $2:3$ (60 and 90 are divisible by $30 \Rightarrow 60 \div 30 = 2$, $90 \div 30 = 3$, Then: $60:90 = 2:3$)
9) $3:8$ (30 and 80 are divisible by $10 \Rightarrow 30 \div 10 = 3$, $80 \div 10 = 8$, Then: $30:80 = 3:8$)
10) $6:13$ (12 and 26 are divisible by $2 \Rightarrow 12 \div 2 = 6, 26 \div 2 = 13$, Then: $12:26 = 6:13$)
11) $9:4$ (63 and 28 are divisible by $7 \Rightarrow 63 \div 7 = 9, 28 \div 7 = 4$, Then: $63:28 = 9:4$)
12) $6:11$ (36 and 66 are divisible by $6 \Rightarrow 36 \div 6 = 6$, $66 \div 6 = 11$, Then: $36:66 = 6:11$)

Proportional Ratios

1) 11 (Use cross multiplication: $\frac{1}{4} = \frac{x}{44} \Rightarrow 1 \times 44 = 4 \times x \Rightarrow 44 = 4x$ Divide both sides by 4 to find x: $x = \frac{44}{4} \Rightarrow x = 11$)

2) 56 (Use cross multiplication: $\frac{1}{7} = \frac{8}{x} \Rightarrow 1 \times x = 7 \times 8 \Rightarrow x = 56$)

3) 49 (Use cross multiplication: $\frac{2}{7} = \frac{14}{x} \Rightarrow 2 \times x = 7 \times 14 \Rightarrow 2x = 98$ Divide both sides by 2 to find x: $x = \frac{98}{2} \Rightarrow x = 49$)

4) 27 (Use cross multiplication: $\frac{3}{10} = \frac{x}{90} \Rightarrow 3 \times 90 = 10 \times x \Rightarrow 270 = 10x$ Divide both sides by 10 to find x: $x = \frac{270}{10} \Rightarrow x = 27$)

5) 52 (Use cross multiplication: $\frac{4}{5} = \frac{x}{65} \Rightarrow 4 \times 65 = 5 \times x \Rightarrow 260 = 5x$ Divide both sides by 5 to find x: $x = \frac{260}{5} \Rightarrow x = 52$)

6) 90 (Use cross multiplication: $\frac{1}{6} = \frac{15}{x} \Rightarrow 1 \times x = 6 \times 15 \Rightarrow x = 90$)

7) 144 (Use cross multiplication: $\frac{4}{9} = \frac{64}{x} \Rightarrow 4 \times x = 9 \times 64 \Rightarrow 4x = 576$ Divide both sides by 4 to find x: $x = \frac{576}{4} \Rightarrow x = 144$)

8) 180 (Use cross multiplication: $\frac{5}{12} = \frac{75}{x} \Rightarrow 5 \times x = 12 \times 75 \Rightarrow 5x = 900$ Divide both sides by 5 to find x: $x = \frac{900}{5} \Rightarrow x = 180$)

Similarity and Ratios

1) 63 (Let's put x for the length of the second rectangle. Since two rectangles are similar, their corresponding sides are in proportion. Write a proportion and solve for the missing number. $\frac{8}{28} = \frac{18}{x} \rightarrow 8x = 18 \times 28 \rightarrow 8x = 504 \rightarrow x = \frac{504}{8} = 63$)

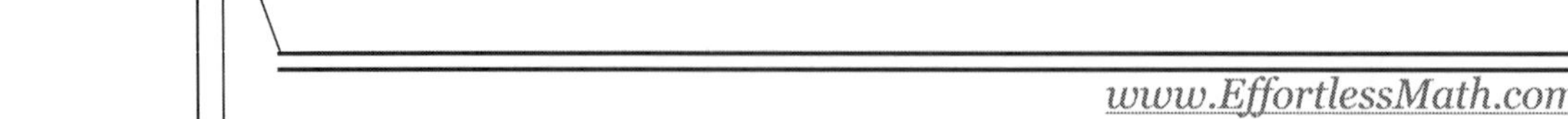

2) 2 (Let's put x for the length of the second rectangle. Since two rectangles are similar, their corresponding sides are in proportion. Write a proportion and solve for the missing number. $\frac{6}{36} = \frac{x}{12} \rightarrow 6 \times 12 = 36x \rightarrow 72 = 36x \rightarrow x = \frac{72}{36} = 2$)

3) 66 (Find the corresponding sides and write a proportion.

$\frac{3}{33} = \frac{6}{x}$. Now, use the cross product to solve for x:

$\frac{3}{33} = \frac{6}{x} \rightarrow 3 \times x = 6 \times 33 \rightarrow 3x = 198$. Divide both sides by 3. Then: $3x = 198 \rightarrow x = \frac{198}{3} \rightarrow x = 66$)

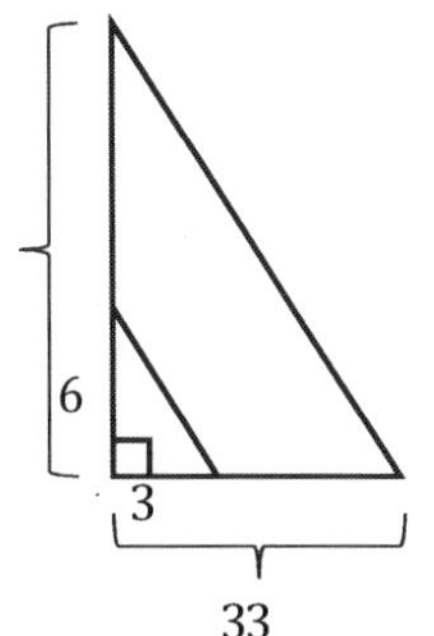

CHAPTER

5 Percentage

Math topics that you'll learn in this chapter:

- ☑ Percent Problems
- ☑ Percent of Increase and Decrease
- ☑ Discount, Tax and Tip
- ☑ Simple Interest

Percent Problems

- Percent is a ratio of a number and 100. It always has the same denominator, 100. The percent symbol is "%".
- Percent means "per 100". So, 20% is $\frac{20}{100}$.
- In each percent problem, we are looking for the base, or the part, or the percent.
- Use these equations to find each missing section in a percent problem:
  - Base = Part ÷ Percent
  - Part = Percent × Base
  - Percent = Part ÷ Base

Examples:

Example 1. What is 20% of 40?

Solution: In this problem, we have the percent (20%) and the base (40) and we are looking for the "part". Use this formula: $Part = Percent \times Base$.
Then: $Part = 20\% \times 40 = \frac{20}{100} \times 40 = 0.20 \times 40 = 8$. The answer: 20% of 40 is 8.

Example 2. 25 is what percent of 500?

Solution: In this problem, we are looking for the percent. Use this equation: $Percent = Part \div Base \rightarrow Percent = 25 \div 500 = 0.05 = 5\%$.
Then: 25 is 5 percent of 500.

Practices:

Solve each problem.

1) 30 is what percent of 40? ____%
2) 22 is what percent of 88? ____%
3) 42 is what percent of 150? ____%
4) 85 is what percent of 340? ____%
5) 84 is 14 percent of what number? ____
6) 72 is 32 percent of what number? ____
7) 12 is 75 percent of what number? ____
8) 64 is 16 percent of what number? ____

Percent of Increase and Decrease

- Percent of change (increase or decrease) is a mathematical concept that represents the degree of change over time.
- To find the percentage of increase or decrease:
  1. New Number – Original Number
  2. (The result ÷ Original Number) × 100
- Or use this formula: $Percent\ of\ change = \frac{new\ number - original\ number}{original\ number} \times 100$
- Note: If your answer is a negative number, then this is a percentage decrease. If it is positive, then this is a percentage increase.

Examples:

Example 1. The price of a shirt increases from \$30 to \$36. What is the percentage increase?

Solution: First, find the difference: $36 - 30 = 6$

Then: $(6 \div 30) \times 100 = \frac{6}{30} \times 100 = 20$. The percentage increase is 20%. It means that the price of the shirt increased by 20%.

Example 2. The price of a table decreased from \$50 to \$35. What is the percent of decrease?

Solution: Use this formula:

$Percent\ of\ change = \frac{new\ number - original\ number}{original\ number} \times 100 =$

$\frac{35-50}{50} \times 100 = \frac{-15}{50} \times 100 = -30$. The percentage decrease is 30. (the negative sign means percentage decrease) Therefore, the price of the table decreased by 30%.

Practices:

Solve each percent of change word problem.

1) John got a raise, and his hourly wage increased from \$16 to \$20. What is the percent increase? _____ %
2) The price of a book increases from \$25 to \$33. What is the percent increase? ___ %
3) 3 of the 15 pieces of a cake have been eaten. . What is the percent decrease of the cake? _____ %

Find more at bit.ly/3pgPQes

Discount, Tax and Tip

- To find the discount: Multiply the regular price by the rate of discount
- To find the selling price: Original price – discount
- To find tax: Multiply the tax rate to the taxable amount (income, property value, etc.)
- To find the tip, multiply the rate by the selling price.

Examples:

Example 1. With a 20% discount, Ella saved \$50 on a dress. What was the original price of the dress?

Solution: let x be the original price of the dress. Then: $20\,\%\ of\ x = 50$. Write an equation and solve for x: $0.20 \times x = 50 \rightarrow x = \frac{50}{0.20} = 250$. The original price of the dress was \$250.

Example 2. Sophia purchased a new computer for a price of \$820 at the Apple Store. What is the total amount her credit card is charged if the sales tax is 5%?

Solution: The taxable amount is \$820, and the tax rate is 5%. Then:

$$Tax = 0.05 \times 820 = 41$$

$Final\ price = Selling\ price + Tax \rightarrow final\ price = \$820 + \$41 = \$861$

Practices:

Find the selling price of each item.

1) Original price of a television: \$450

   Tax: 8%, Selling price: \$_______

2) Original price of a sculpture: \$700

   Tax: 11%, Selling price: \$_______

3) Original price of a stove: \$1,000

   Tax: 7.5%, Selling price: \$_______

4) Original price of the sunglasses:\$240

   Discount: 15%, Selling price:\$_____

5) Original price of a Table: \$380

   Discount: 25%, Selling price:\$_____

6) Original price of a car: \$25,000

   Discount: 5%, Selling price:\$______

Simple Interest

- Simple Interest: The charge for borrowing money or the return for lending it.
- Simple interest is calculated on the initial amount (principal).
- To solve a simple interest problem, use this formula:

$$Interest = principal \times rate \times time \qquad (I = p \times r \times t = prt)$$

Examples:

Example 1. Find simple interest for \$200 investment at 5% for 3 years.

Solution: Use Interest formula:

$I = prt$ ($p = \$200$, $r = 5\% = \frac{5}{100} = 0.05$ and $t = 3$)

Then: $I = 200 \times 0.05 \times 3 = \$30$

Example 2. Find simple interest for \$1,200 at 8% for 6 years.

Solution: Use Interest formula:

$I = prt$ ($p = \$1{,}200$, $r = 8\% = \frac{8}{100} = 0.08$ and $t = 6$)

Then: $I = 1{,}200 \times 0.08 \times 6 = \$576$

Practices:

✍ ***Determine the simple interest for these loans.***

1) \$700 at 3% for 2 years. $ ________
2) \$1,250 at 6.5% for 4 years. $ ________
3) \$13,000 at 14% for 6 months. $ ________
4) \$850 at 10% for 7 years. $ ________
5) \$11,300 at 4% for 9 months. $ ________
6) \$2,400 at 5% for 5 years. $ ________
7) \$950 at 6% for 8 years. $ ________
8) \$3,800 at 2% for 3 years. $ ________

Chapter 5: Answers

Percent Problems

1) 75% (In this problem, we are looking for the percent. Use this equation: $Percent = Part \div Base \rightarrow Percent = 30 \div 40 = 0.75 = 75\%$. Then: 30 is 75 percent of 40)

2) 25% (In this problem, we are looking for the percent. Use this equation: $Percent = Part \div Base \rightarrow Percent = 22 \div 88 = 0.25 = 25\%$. Then: 22 is 25 percent of 88)

3) 28% (In this problem, we are looking for the percent. Use this equation: $Percent = Part \div Base \rightarrow Percent = 42 \div 150 = 0.28 = 28\%$. Then: 42 is 28 percent of 150)

4) 25% (In this problem, we are looking for the percent. Use this equation: $Percent = Part \div Base \rightarrow Percent = 85 \div 340 = 0.25 = 25\%$. Then: 85 is 25 percent of 340)

5) 600 (In this problem, we are looking for the base. Use this equation: $Base = Part \div Percent \rightarrow Base = 84 \div 14\% = 84 \div 0.14 = 600$. Then: 84 is 14 percent of 600)

6) 225 (In this problem, we are looking for the base. Use this equation: $Base = Part \div Percent \rightarrow Base = 72 \div 32\% = 72 \div 0.32 = 225$. Then: 72 is 32 percent of 225)

7) 16 (In this problem, we are looking for the base. Use this equation: $Base = Part \div Percent \rightarrow Base = 12 \div 75\% = 12 \div 0.75 = 16$. Then: 12 is 75 percent of 16)

8) 400 (In this problem, we are looking for the base. Use this equation: $Base = Part \div Percent \rightarrow Base = 64 \div 16\% = 12 \div 0.16 = 400$. Then: 64 is 16 percent of 400)

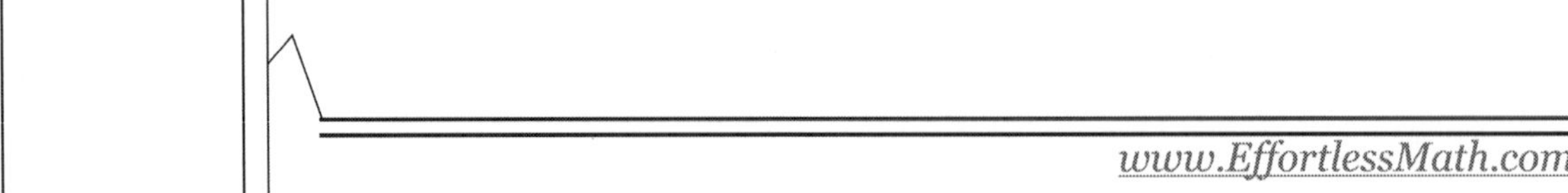

Percent of Increase and Decrease

1) 25 (First, find the difference: $20 - 16 = 4$. Then: $(4 \div 16) \times 100 = \frac{4}{16} \times 100 = 25$. The percentage increase is 25%. It means that the price of the shirt increased by 25%.)

2) 32 (First, find the difference: $33 - 25 = 8$. Then: $(8 \div 25) \times 100 = \frac{8}{25} \times 100 = 32$. The percentage increase is 32%. It means that the price of the shirt increased by 32%.)

3) 20% (Use this formula: $Percent\ of\ change = \frac{new\ number - original\ number}{original\ number} \times 100 = \frac{12-15}{15} \times 100 = \frac{-3}{15} \times 100 = -20$. The percentage decrease is 20. (the negative sign means percentage decrease) Therefore, the price of the table decreased by 20%.)

Discount, Tax and Tip

1) \$486.00 (The taxable amount is \$450, and the tax rate is 8%. Then: $Tax = 0.08 \times 450 = 36.\ Final\ price = Selling\ price + Tax \rightarrow final\ price = \$450 + \$36 = \$486$)

2) \$777.00 (The taxable amount is \$700, and the tax rate is 11%. Then: $Tax = 0.11 \times 700 = 77.\ Final\ price = Selling\ price + Tax \rightarrow final\ price = \$700 + \$77 = \$777$)

3) \$1,075.00 (The taxable amount is \$1,000, and the tax rate is 7.5%. Then: $Tax = 0.075 \times 1{,}000 = 75.\ Final\ price = Selling\ price + Tax \rightarrow final\ price = \$1{,}000 + \$75 = \$1{,}075$)

4) \$204.00 (Original price of the dress is \$240. Then: $15\%\ of\ 240 = 36$. $Final\ price = Selling\ price - Discount \rightarrow final\ price = \$240 - \$36 = \$204$)

5) \$285.00 (Original price of the dress is \$380. Then: $25\%\ of\ 380 = 95$. $Final\ price = Selling\ price - Discount \rightarrow final\ price = \$380 - \$95 = \$285$)

6) \$23,750.00 (Original price of the dress is \$25,000. Then: $5\%\ of\ 25,000 = 1,250$. $Final\ price = Selling\ price - Discount \rightarrow final\ price = \$25,000 - \$1,250 = \$23,750.00$)

Simple Interest

1) \$42.00 (Use Interest formula: $I = prt$ ($p = \$700$, $r = 3\% = \frac{3}{100} = 0.03$ and $t = 2$)Then: $I = 700 \times 0.03 \times 2 = \$42$)

2) \$325.00 (Use Interest formula: $I = prt$ ($p = \$1,250$, $r = 6.5\% = \frac{6.5}{100} = 0.065$ and $t = 4$)Then: $I = 1,250 \times 0.065 \times 4 = \$325$)

3) \$910.00 (Use Interest formula: $I = prt.\ p = \$13,000,\ r = 14\% = 0.14$ and $t = 0.5$ (6 months is half year). Then: $I = 13,000 \times 0.14 \times 0.5 = \$910$)

4) \$595.00 (Use Interest formula: $I = prt$ ($p = \$850$, $r = 10\% = \frac{10}{100} = 0.1$ and $t = 7$)Then: $I = 850 \times 0.1 \times 7 = \$595$)

5) \$339.00 (Use Interest formula: $I = prt.\ p = \$11,300,\ r = 4\% = 0.04$ and $t = 0.75$ (9 months is $\frac{3}{4}$ year). Then: $I = 11,300 \times 0.04 \times 0.75 = \$339$)

6) \$600.00 (Use Interest formula: $I = prt$ ($p = \$2,400$, $r = 5\% = \frac{5}{100} = 0.05$ and $t = 5$)Then: $I = 2,400 \times 0.05 \times 5 = \$600$)

7) \$456.00 (Use Interest formula: $I = prt$ ($p = \$950$, $r = 6\% = \frac{6}{100} = 0.06$ and $t = 8$)Then: $I = 950 \times 0.06 \times 8 = \$456$)

8) \$228.00 (Use Interest formula: $I = prt$ ($p = \$3,800$, $r = 2\% = \frac{2}{100} = 0.02$ and $t = 3$)Then: $I = 3,800 \times 0.02 \times 3 = \$228$)

CHAPTER

6 Exponents and Variables

Math topics that you'll learn in this chapter:

- ☑ Multiplication Property of Exponents
- ☑ Division Property of Exponents
- ☑ Powers of Products and Quotients
- ☑ Zero and Negative Exponents
- ☑ Negative Exponents and Negative Bases
- ☑ Scientific Notation
- ☑ Radicals

Multiplication Property of Exponents

- Exponents are shorthand for repeated multiplication of the same number by itself. For example, instead of 2×2, we can write 2^2. For $3 \times 3 \times 3 \times 3$, we can write 3^4
- In algebra, a variable is a letter used to stand for a number. The most common letters are: x, y, z, a, b, c, m, and n.
- Exponent's rules: $x^a \times x^b = x^{a+b}$, $\frac{x^a}{x^b} = x^{a-b}$

$(x^a)^b = x^{a \times b}$ $\quad$ $(xy)^a = x^a \times y^a$ $\quad$ $\left(\frac{a}{b}\right)^c = \frac{a^c}{b^c}$

Examples:

Example 1. Multiply. $2x^2 \times 3x^4$

Solution: Use Exponent's rules: $x^a \times x^b = x^{a+b} \rightarrow x^2 \times x^4 = x^{2+4} = x^6$
Then: $2x^2 \times 3x^4 = 6x^6$

Example 2. Simplify. $(x^4y^2)^2$

Solution: Use Exponent's rules: $(x^a)^b = x^{a \times b}$.
Then: $(x^4y^2)^2 = x^{4\times2}y^{2\times2} = x^8y^4$

Practices:

Simplify and write the answer in exponential form.

1) $x^3 \times 4x =$
2) $x \times 3x^5 =$
3) $3x^3 \times 6x^4 =$
4) $7yx^4 \times 4x =$
5) $2x^5 \times y^3x^3 =$
6) $y^3x^4 \times y^8x^6 =$
7) $6yx^5 \times 3x^7y^3 =$
8) $9x^6 \times 4x^8y^2 =$

Division Property of Exponents

For division of exponents use following formulas:

- $\frac{x^a}{x^b} = x^{a-b}\ (x \neq 0)$
- $\frac{x^a}{x^b} = \frac{1}{x^{b-a}},\ (x \neq 0)$
- $\frac{1}{x^b} = x^{-b}$

Examples:

Example 1. Simplify. $\frac{16x^3y}{2xy^2} =$

Solution: First, cancel the common factor: $2 \rightarrow \frac{16x^3y}{2xy^2} = \frac{8x^3y}{xy^2}$

Use Exponent's rules: $\frac{x^a}{x^b} = x^{a-b} \rightarrow \frac{x^3}{x} = x^{3-1} = x^2$ and $\frac{x^a}{x^b} = \frac{1}{x^{b-a}} \rightarrow \frac{y}{y^2} = \frac{1}{y^{2-1}} = \frac{1}{y}$

Then: $\frac{16x^3y}{2xy^2} = \frac{8x^2}{y}$

Example 2. Simplify. $\frac{7x^4y^2}{28x^3y} =$

Solution: First, cancel the common factor: $7 \rightarrow \frac{x^4y^2}{4x^3y}$

Use Exponent's rules: $\frac{x^a}{x^b} = x^{a-b} \rightarrow \frac{x^4}{x^3} = x^{4-3} = x$ and $\frac{y^2}{y} = y$

Then: $\frac{7x^4y^2}{28x^3y} = \frac{xy}{4}$

Practices:

Simplify.

1) $\frac{4^5 \times 4^8}{4^7 \times 4^4} =$

2) $\frac{8x^8}{24x} =$

3) $\frac{8y^5}{3y^{12}} =$

4) $\frac{8x^4}{24x^9} =$

5) $\frac{21x^{32}}{7y^{21}} =$

6) $\frac{49xy^{13}}{56y^7} =$

7) $\frac{5x^7}{16x} =$

8) $\frac{63x^6y^4}{9x^9} =$

9) $\frac{15x^5y^6}{20x^{11}y^9} =$

Find more at bit.ly/37JAclZ

Powers of Products and Quotients

- For any nonzero numbers a and b and any integer x, $(ab)^x = a^x \times b^x$ and $\left(\frac{a}{b}\right)^c = \frac{a^c}{b^c}$

Examples:

Example 1. Simplify. $(3x^3y^2)^2$

Solution: Use Exponent's rules: $(x^a)^b = x^{a\times b}$

$(3x^3y^2)^2 = (3)^2(x^3)^2(y^2)^2 = 9x^{3\times2}y^{2\times2} = 9x^6y^4$

Example 2. Simplify. $\left(\frac{2x^3}{3x^2}\right)^2$

Solution: First, cancel the common factor: $x \rightarrow \left(\frac{2x^3}{3x^2}\right) = \left(\frac{2x}{3}\right)^2$

Use Exponent's rules: $\left(\frac{a}{b}\right)^c = \frac{a^c}{b^c}$, Then: $\left(\frac{2x}{3}\right)^2 = \frac{(2x)^2}{(3)^2} = \frac{4x^2}{9}$

Practices:

Simplify.

1) $(5x^3y^4)^3 =$

2) $(4x^6 \times 3x)^4 =$

3) $(3x^3y^7)^4 =$

4) $(5x \times 5y^5)^3 =$

5) $\left(\frac{9x^3}{x^4}\right)^3 =$

6) $\left(\frac{x^3y^5}{x^6y^2}\right)^6 =$

7) $\left(\frac{xy^4}{x^5y^3}\right)^4 =$

8) $\left(\frac{2xy^3}{x^4}\right)^5 =$

Find more at bit.ly/34CgPJm

Zero and Negative Exponents

- Zero-Exponent Rule: $a^0 = 1$, this means that anything raised to the zero power is 1. For example: $(5xy)^0 = 1$ (number zero is an exception: $0^0 = 0$)
- A negative exponent simply means that the base is on the wrong side of the fraction line, so you need to flip the base to the other side. For instance, "x^{-2}" (pronounced as "ecks to the minus two") just means "x^2" but underneath, as in $\frac{1}{x^2}$.

Examples:

Example 1. Evaluate. $\left(\frac{4}{5}\right)^{-2} =$

Solution: Use negative exponent's rule: $\left(\frac{x^a}{x^b}\right)^{-2} = \left(\frac{x^b}{x^a}\right)^2 \rightarrow \left(\frac{4}{5}\right)^{-2} = \left(\frac{5}{4}\right)^2$
Then: $\left(\frac{5}{4}\right)^2 = \frac{5^2}{4^2} = \frac{25}{16}$

Example 2. Evaluate. $\left(\frac{a}{b}\right)^0 =$

Solution: Use zero-exponent Rule: $a^0 = 1$
Then: $\left(\frac{a}{b}\right)^0 = 1$

Practices:

Evaluate the following expressions.

1) $3^{-4} =$
2) $5^{-2} =$
3) $4^{-3} =$
4) $10^{-5} =$
5) $10^{-7} =$
6) $\left(\frac{1}{4}\right)^{-2} =$
7) $\left(\frac{3}{2}\right)^{-2} =$
8) $\left(\frac{1}{2}\right)^0 =$

Negative Exponents and Negative Bases

- A negative exponent is the reciprocal of that number with a positive exponent. $(3)^{-2} = \frac{1}{3^2}$
- To simplify a negative exponent, make the power positive!
- The parenthesis is important! -5^{-2} is not the same as $(-5)^{-2}$

$$-5^{-2} = -\frac{1}{5^2} \text{ and } (-5)^{-2} = +\frac{1}{5^2}$$

Examples:

Example 1. Simplify. $\left(\frac{2a}{3c}\right)^{-2} =$

Solution: Use negative exponent's rule: $\left(\frac{x^a}{x^b}\right)^{-2} = \left(\frac{x^b}{x^a}\right)^2 \rightarrow \left(\frac{2a}{3c}\right)^{-2} = \left(\frac{3c}{2a}\right)^2$

Now use exponent's rule: $\left(\frac{a}{b}\right)^c = \frac{a^c}{b^c} \rightarrow \left(\frac{3c}{2a}\right)^2 = \frac{3^2c^2}{2^2a^2}$

Then: $\frac{3^2c^2}{2^2a^2} = \frac{9c^2}{4a^2}$

Example 2. Simplify. $\left(\frac{x}{4y}\right)^{-3} =$

Solution: Use negative exponent's rule: $\left(\frac{x^a}{x^b}\right)^{-3} = \left(\frac{x^b}{x^a}\right)^3 \rightarrow \left(\frac{x}{4y}\right)^{-3} = \left(\frac{4y}{x}\right)^3$

Now use exponent's rule: $\left(\frac{a}{b}\right)^c = \frac{a^c}{b^c} \rightarrow \left(\frac{4y}{x}\right)^3 = \frac{4^3y^3}{x^3} = \frac{64y^3}{x^3}$

Practices:

Simplify.

1) $-7x^{-3}y^{-2} =$

2) $17x^{-1}y^{-7} =$

3) $8a^{-5}b^{-3} =$

4) $-10a^{-2}b^{-9} =$

5) $-\frac{16}{x^{-3}} =$

6) $\frac{12b}{-6c^{-2}} =$

Scientific Notation

- Scientific notation is used to write very big or very small numbers in decimal form.
- In scientific notation, all numbers are written in the form of: $m \times 10^n$, where m is greater than 1 and less than 10.
- To convert a number from scientific notation to standard form, move the decimal point to the left (if the exponent of ten is a negative number), or to the right (if the exponent is positive).

Examples:

Example 1. Write 0.00024 in scientific notation.

Solution: First, move the decimal point to the right so you have a number between 1 and 10. That number is 2.4. Now, determine how many places the decimal moved in step 1 by the power of 10. We moved the decimal point 4 digits to the right. Then: $10^{-4} \rightarrow$ When the decimal moved to the right, the exponent is negative. Then: $0.00024 = 2.4 \times 10^{-4}$

Example 2. Write 3.8×10^{-5} in standard notation.

Solution: The exponent is negative 5. Then, move the decimal point to the left five digits. (remember $3.8 = 0000003.8$) When the decimal moved to the right, the exponent is negative. Then: $3.8 \times 10^{-5} = 0.000038$

Practices:

Write each number in scientific notation.

1) $0.00004869 =$
2) $0.00289 =$
3) $67{,}000{,}000 =$
4) $943{,}000 =$

Write each number in standard notation.

5) $5 \times 10^{-5} =$
6) $3.2 \times 10^{-4} =$
7) $3.4 \times 10^{-6} =$
8) $6.8 \times 10^{-7} =$

Find more at bit.ly/3nOwJYP

Radicals

- If n is a positive integer and x is a real number, then: $\sqrt[n]{x} = x^{\frac{1}{n}}$,

$$\sqrt[n]{xy} = x^{\frac{1}{n}} \times y^{\frac{1}{n}},\ \sqrt[n]{\frac{x}{y}} = \frac{x^{\frac{1}{n}}}{y^{\frac{1}{n}}},\ \text{and } \sqrt[n]{x} \times \sqrt[n]{y} = \sqrt[n]{xy}$$

- A square root of x is a number r whose square is: $r^2 = x$ (r is a square root of x)
- To add and subtract radicals, we need to have the same values under the radical. For example: $\sqrt{3} + \sqrt{3} = 2\sqrt{3}$, $3\sqrt{5} - \sqrt{5} = 2\sqrt{5}$

Examples:

Example 1. Find the square root of $\sqrt{121}$.

Solution: First, factor the number: $121 = 11^2$, Then: $\sqrt{121} = \sqrt{11^2}$,
Now use radical rule: $\sqrt[n]{a^n} = a$. Then: $\sqrt{121} = \sqrt{11^2} = 11$

Example 2. Evaluate. $\sqrt{4} \times \sqrt{16} =$

Solution: Find the values of $\sqrt{4}$ and $\sqrt{16}$. Then: $\sqrt{4} \times \sqrt{16} = 2 \times 4 = 8$

Example 3. Solve. $5\sqrt{2} + 9\sqrt{2}$.

Solution: Since we have the same values under the radical, we can add these two radicals: $5\sqrt{2} + 9\sqrt{2} = 14\sqrt{2}$

Practices:

Evaluate.

1) $\sqrt{36} \times \sqrt{81} =$ ________
2) $\sqrt{3} \times \sqrt{27} =$ ________
3) $\sqrt{27} \times \sqrt{27} =$ ________
4) $\sqrt{125} + \sqrt{125} =$ ________
5) $6\sqrt{7} - 3\sqrt{7} =$ ________
6) $4\sqrt{10} \times 2\sqrt{10} =$ ________
7) $7\sqrt{2} \times 3\sqrt{2} =$ ________
8) $7\sqrt{5} - \sqrt{20} =$ ________

Chapter 6: Answers

Multiplication Property of Exponents

1) $4x^4$ (Use Exponent's rules: $x^a \times x^b = x^{a+b} \rightarrow x^3 \times x = x^{3+1} = x^4$.

   Then: $x^3 \times 4x = 4x^4$)

2) $3x^6$ (Use Exponent's rules: $x^a \times x^b = x^{a+b} \rightarrow x \times x^5 = x^{1+5} = x^6$.

   Then: $x \times 3x^5 = 3x^6$)

3) $18x^7$ (Use Exponent's rules: $x^a \times x^b = x^{a+b} \rightarrow x^3 \times x^4 = x^{3+4} = x^7$.

   Then: $3x^3 \times 6x^4 = 18x^7$)

4) $28yx^5$ (Use Exponent's rules: $x^a \times x^b = x^{a+b} \rightarrow yx^4 \times x = yx^{4+1} = yx^5$.

   Then: $7yx^4 \times 4x = 28yx^5$)

5) $2x^8y^3$ (Use Exponent's rules: $x^a \times x^b = x^{a+b} \rightarrow x^5 \times y^3x^3 = x^{5+3}y^3 = x^8y^3$. Then: $2x^5 \times y^3x^3 = 2x^8y^3$)

6) $y^{11}x^{10}$ (Use Exponent's rules: $x^a \times x^b = x^{a+b} \rightarrow y^3x^4 \times y^8x^6 = y^{3+8}x^{4+6} = y^{11}x^{10}$)

7) $18x^{12}y^4$ (Use Exponent's rules: $x^a \times x^b = x^{a+b} \rightarrow yx^5 \times x^7y^3 = x^{5+7}y^{1+3} = x^{12}y^4$. Then: $6yx^5 \times 3x^7y^3 = 18x^{12}y^4$)

8) $36x^{14}y^2$ (Use Exponent's rules: $x^a \times x^b = x^{a+b} \rightarrow x^6 \times x^8y^2 = x^{6+8}y^2 = x^{14}y^2$. Then: $9x^6 \times 4x^8y^2 = 36x^{14}y^2$)

Division Property of Exponents

1) 16 (Use Exponent's rules: $x^a \times x^b = x^{a+b} \rightarrow 4^5 \times 4^8 = 4^{5+8} = 4^{13}$ and $4^7 \times 4^4 = 4^{7+4} = 4^{11}$. Use Exponent's rules: $\frac{x^a}{x^b} = x^{a-b} \rightarrow \frac{4^{13}}{4^{11}} = 4^{13-11} = 4^2 = 16$)

2) $\frac{x^7}{3}$ (Cancel the common factor: $8 \rightarrow \frac{8x^8}{24x} = \frac{x^8}{3x}$. Use Exponent's rules: $\frac{x^a}{x^b} = x^{a-b} \rightarrow \frac{x^8}{x} = x^{8-1} = x^7$. Then: $\frac{8x^8}{24x} = \frac{x^7}{3}$)

3) $\frac{8}{3y^7}$ (Use Exponent's rules: $\frac{x^a}{x^b} = \frac{1}{x^{b-a}} \rightarrow \frac{y^5}{y^{12}} = \frac{1}{y^{12-5}} = \frac{1}{y^7}$. Then: $\frac{8y^5}{3y^{12}} = \frac{8}{3y^7}$)

4) $\frac{1}{3x^5}$ (Cancel the common factor: $8 \rightarrow \frac{8x^4}{24x^9} = \frac{x^4}{3x^9}$. Use Exponent's rules: $\frac{x^a}{x^b} = \frac{1}{x^{b-a}} \rightarrow \frac{x^4}{x^9} = \frac{1}{x^{9-4}} = \frac{1}{x^5}$. Then: $\frac{8x^4}{24x^9} = \frac{1}{3x^5}$)

5) $\frac{3x^{32}}{y^{21}}$ (Cancel the common factor: $7 \rightarrow \frac{21x^{32}}{7y^{21}} = \frac{3x^{32}}{y^{21}}$)

6) $\frac{7xy^6}{8}$ (Cancel the common factor: $7 \rightarrow \frac{49xy^{13}}{56y^7} = \frac{7xy^{13}}{8y^7}$. Use Exponent's rules: $\frac{x^a}{x^b} = x^{a-b} \rightarrow \frac{y^{13}}{y^7} = y^{13-7} = y^6$. Then: $\frac{49xy^{13}}{57y^7} = \frac{7xy^6}{8}$)

7) $\frac{5x^6}{16}$ (Use Exponent's rules: $\frac{x^a}{x^b} = x^{a-b} \rightarrow \frac{x^7}{x} = x^{7-1} = x^6$. Then: $\frac{5x^7}{16x} = \frac{5x^6}{16}$)

8) $\frac{7y^4}{x^3}$ (Cancel the common factor: $9 \rightarrow \frac{63x^6y^4}{9x^9} = \frac{7x^6y^4}{x^9}$. Use Exponent's rules: $\frac{x^a}{x^b} = \frac{1}{x^{b-a}} \rightarrow \frac{x^6}{x^9} = \frac{1}{x^{9-6}} = \frac{1}{x^3}$. Then: $\frac{63x^6y^4}{9x^9} = \frac{7y^4}{x^3}$)

9) $\frac{3}{4x^6y^3}$ (Cancel the common factor: $5 \rightarrow \frac{15x^5y^6}{20x^{11}y^9} = \frac{3x^5y^6}{4x^{11}y^9}$. Use Exponent's rules: $\frac{x^a}{x^b} = \frac{1}{x^{b-a}} \rightarrow \frac{x^5}{x^{11}} = \frac{1}{x^{11-5}} = \frac{1}{x^6}$ and $\frac{y^6}{y^9} = \frac{1}{y^{9-6}} = \frac{1}{y^3}$. Then: $\frac{15x^5y^6}{20x^{11}y^9} = \frac{3}{4x^6y^3}$)

Powers of Products and Quotients

1) $125x^9y^{12}$ (Use Exponent's rules: $(x^a)^b = x^{a\times b} \rightarrow (5x^3y^4)^3 = (5)^3(x^3)^3(y^4)^3 = 125x^{3\times3}y^{4\times3} = 125x^9y^{12}$)

2) $20{,}736x^{28}$ (Use Exponent's rules: $(x^a)^b = x^{a\times b} \rightarrow (4x^6 \times 3x)^4 = (4)^4(x^6)^4(3)^4(x)^4 = 256x^{6\times4}81x^4 = 256x^{24}81x^4 = 20{,}736x^{28}$)

3) $81x^{12}y^{28}$ (Use Exponent's rules: $(x^a)^b = x^{a\times b} \rightarrow (3x^3y^7)^4 = (3)^4(x^3)^4(y^7)^4 = 81x^{12}y^{28}$)

4) $15{,}625x^3y^{15}$ (Use Exponent's rules: $(x^a)^b = x^{a\times b} \rightarrow (5x \times 5y^5)^3 = (5)^3(x)^3 \times (5)^3\left(y^5\right)^3 = 125x^3 \times 125y^{5\times3} = 125x^3 \times 125y^{15} = 15{,}625x^3y^{15}$)

5) $\frac{729}{x^3}$ (First, cancel the common factor: $x \rightarrow \left(\frac{9x^3}{x^4}\right)^3 = \left(\frac{9}{x}\right)^3$. Use Exponent's rules: $\left(\frac{a}{b}\right)^c = \frac{a^c}{b^c}$, Then: $\left(\frac{9}{x}\right)^3 = \frac{(9)^3}{(x)^3} = \frac{729}{x^3}$)

6) $\frac{y^{18}}{x^{18}}$ (First, cancel the common factor: $x \ and \ y \rightarrow \left(\frac{x^3y^5}{x^6y^2}\right)^6 = \left(\frac{y^3}{x^3}\right)^6$. Use Exponent's rules: $\left(\frac{a}{b}\right)^c = \frac{a^c}{b^c}$, Then: $\left(\frac{y^3}{x^3}\right)^6 = \frac{(y^3)^6}{(x^3)^6} = \frac{y^{3\times6}}{x^{3\times6}} = \frac{y^{18}}{x^{18}}$)

7) $\frac{y^4}{x^{16}}$ (First, cancel the common factor: x and $y \rightarrow \left(\frac{xy^4}{x^5y^3}\right)^4 = \left(\frac{y}{x^4}\right)^4$. Use Exponent's rules: $\left(\frac{a}{b}\right)^c = \frac{a^c}{b^c}$, Then: $\left(\frac{y}{x^4}\right)^4 = \frac{(y)^4}{(x^4)^4} = \frac{y^4}{x^{4\times4}} = \frac{y^4}{x^{16}}$)

8) $\frac{32y^{15}}{x^{15}}$ (First, cancel the common factor: $x \rightarrow \left(\frac{2xy^3}{x^4}\right)^5 = \left(\frac{2y^3}{x^3}\right)^5$. Use Exponent's rules: $\left(\frac{a}{b}\right)^c = \frac{a^c}{b^c}$, Then: $\left(\frac{2y^3}{x^3}\right)^5 = \frac{(2)^5(y^3)^5}{(x^3)^5} = \frac{32y^{3\times5}}{x^{3\times5}} = \frac{32y^{15}}{x^{15}}$)

Zero and Negative Exponents

1) $\frac{1}{81}$ (Use negative exponent's rule: $\left(\frac{x^a}{x^b}\right)^{-2} = \left(\frac{x^b}{x^a}\right)^2 \rightarrow 3^{-4} = \left(\frac{1}{3}\right)^4$. Then: $\left(\frac{1}{3}\right)^4 = \frac{1^4}{3^4} = \frac{1}{81}$)

2) $\frac{1}{25}$ (Use negative exponent's rule: $\left(\frac{x^a}{x^b}\right)^{-2} = \left(\frac{x^b}{x^a}\right)^2 \rightarrow 5^{-2} = \left(\frac{1}{5}\right)^2$. Then: $\left(\frac{1}{5}\right)^2 = \frac{1^2}{5^2} = \frac{1}{25}$)

3) $\frac{1}{64}$ (Use negative exponent's rule: $\left(\frac{x^a}{x^b}\right)^{-2} = \left(\frac{x^b}{x^a}\right)^2 \rightarrow 4^{-3} = \left(\frac{1}{4}\right)^3$. Then: $\left(\frac{1}{4}\right)^3 = \frac{1^3}{4^3} = \frac{1}{64}$)

4) $\frac{1}{100{,}000}$ (Use negative exponent's rule: $\left(\frac{x^a}{x^b}\right)^{-2} = \left(\frac{x^b}{x^a}\right)^2 \rightarrow 10^{-5} = \left(\frac{1}{10}\right)^5$. Then: $\left(\frac{1}{10}\right)^5 = \frac{1^5}{10^5} = \frac{1}{100{,}000}$)

5) $\frac{1}{10,000,000}$ (Use negative exponent's rule: $\left(\frac{x^a}{x^b}\right)^{-2} = \left(\frac{x^b}{x^a}\right)^2 \rightarrow 10^{-7} = \left(\frac{1}{10}\right)^7$. Then: $\left(\frac{1}{10}\right)^7 = \frac{1^7}{10^7} = \frac{1}{10,000,000}$)

6) 16 (Use negative exponent's rule: $\left(\frac{x^a}{x^b}\right)^{-2} = \left(\frac{x^b}{x^a}\right)^2 \rightarrow \left(\frac{1}{4}\right)^{-2} = \left(\frac{4}{1}\right)^2$. Then: $\left(\frac{4}{1}\right)^2 = \frac{4^2}{1^2} = \frac{16}{1} = 16$)

7) $\frac{4}{9}$ (Use negative exponent's rule: $\left(\frac{x^a}{x^b}\right)^{-2} = \left(\frac{x^b}{x^a}\right)^2 \rightarrow \left(\frac{3}{2}\right)^{-2} = \left(\frac{2}{3}\right)^2$. Then: $\left(\frac{2}{3}\right)^2 = \frac{2^2}{3^2} = \frac{4}{9}$)

8) 1 (Use zero-exponent Rule: $a^0 = 1$. Then: $\left(\frac{a}{b}\right)^0 = 1 \rightarrow \left(\frac{1}{2}\right)^0 = 1$)

Negative Exponents and Negative Bases

1) $-\frac{7}{x^3y^2}$ (Use negative exponent's rule: $\left(\frac{x^a}{x^b}\right)^{-2} = \left(\frac{x^b}{x^a}\right)^2 \rightarrow -7x^{-3}y^{-2} = -7\left(\frac{1}{x}\right)^3\left(\frac{1}{y}\right)^2$. Now use exponent's rule: $\left(\frac{a}{b}\right)^c = \frac{a^c}{b^c} \rightarrow -7\left(\frac{1}{x}\right)^3\left(\frac{1}{y}\right)^2 = -7\frac{1^2\times1^2}{x^3y^2} \rightarrow -\frac{7}{x^3y^2}$)

2) $\frac{17}{xy^7}$ (Use negative exponent's rule: $\left(\frac{x^a}{x^b}\right)^{-2} = \left(\frac{x^b}{x^a}\right)^2 \rightarrow 17x^{-1}y^{-7} = 17\frac{1}{x}\left(\frac{1}{y}\right)^7$. Now use exponent's rule: $\left(\frac{a}{b}\right)^c = \frac{a^c}{b^c} \rightarrow 17\frac{1}{x}\left(\frac{1}{y}\right)^7 = 17\frac{1\times1^7}{xy^7} \rightarrow \frac{17}{xy^7}$

3) $\frac{8}{a^5b^3}$ (Use negative exponent's rule: $\left(\frac{x^a}{x^b}\right)^{-2} = \left(\frac{x^b}{x^a}\right)^2 \rightarrow 8a^{-5}b^{-3} = 8\left(\frac{1}{a}\right)^5\left(\frac{1}{b}\right)^3$. Now use exponent's rule: $\left(\frac{a}{b}\right)^c = \frac{a^c}{b^c} \rightarrow 8\left(\frac{1}{a}\right)^5\left(\frac{1}{b}\right)^3 = 8\frac{1^5\times1^3}{a^5b^3} \rightarrow \frac{8}{a^5b^3}$)

4) $-\frac{10}{a^2b^9}$ (Use negative exponent's rule: $\left(\frac{x^a}{x^b}\right)^{-2} = \left(\frac{x^b}{x^a}\right)^2 \rightarrow -10a^{-2}b^{-9} =$ $-10\left(\frac{1}{a}\right)^2\left(\frac{1}{b}\right)^9$. Now use exponent's rule: $\left(\frac{a}{b}\right)^c = \frac{a^c}{b^c} \rightarrow -10\left(\frac{1}{a}\right)^2\left(\frac{1}{b}\right)^9 =$ $-10\frac{1^2 \times 1^9}{x^2y^9} \rightarrow -\frac{10}{a^2b^9}$)

5) $-16x^3$ (Use negative exponent's rule: $\left(\frac{x^a}{x^b}\right)^{-2} = \left(\frac{x^b}{x^a}\right)^2 \rightarrow -\frac{16}{x^{-3}} = -\frac{16}{\left(\frac{1}{x}\right)^3}$. Now use exponent's rule: $\left(\frac{a}{b}\right)^c = \frac{a^c}{b^c} \rightarrow -\frac{16}{\left(\frac{1}{x}\right)^3} = \frac{16}{\frac{1^3}{x^3}} \rightarrow -\frac{16}{\frac{1}{x^3}} \rightarrow -16x^3$)

6) $-2bc^2$ (First simplify fraction: $\frac{12b}{-6c^{-2}} = -\frac{2b}{c^{-2}}$. Then use negative exponent's rule: $\left(\frac{x^a}{x^b}\right)^{-2} = \left(\frac{x^b}{x^a}\right)^2 \rightarrow -\frac{2b}{c^{-2}} = -\frac{2b}{\left(\frac{1}{c}\right)^2}$. Now use exponent's rule: $\left(\frac{a}{b}\right)^c = \frac{a^c}{b^c} \rightarrow -\frac{2b}{\left(\frac{1}{c}\right)^2} = \frac{2b}{\frac{1^2}{c^2}} \rightarrow -\frac{2b}{\frac{1}{c^2}} \rightarrow -2bc^2$)

Scientific Notation

1) 4.869×10^{-5} (First, move the decimal point to the right so you have a number between 1 and 10. That number is 4.869. Now, determine how many places the decimal moved in step 1 by the power of 10. We moved the decimal point 5 digits to the right. Then: 10^{-5} → When the decimal moved to the right, the exponent is negative. Then: $0.00004869 = 4.869 \times 10^{-5}$)

2) 2.89×10^{-3} (Move the decimal point to the right so you have a number between 1 and 10. That number is 2.89. Now, determine how many places the decimal moved in step 1 by the power of 10. We moved the decimal point 3 digits to the right. Then: 10^{-3} → When the decimal moved to the right, the exponent is negative. Then: $0.00289 = 2.89 \times 10^{-3}$)

3) 6.7×10^7 (Move the decimal point to the left so you have a number between 1 and 10. That number is 6.7. Now, determine how many places the decimal moved in step 1 by the power of 10. We moved the decimal point 7 digits to the right. Then: 10^7 → When the decimal moved to the left, the exponent is positive. Then: $67{,}000{,}000 = 6.7 \times 10^7$)

4) 9.43×10^5 (Move the decimal point to the left so you have a number between 1 and 10. That number is 9.43. Now, determine how many places the decimal moved in step 1 by the power of 10. We moved the decimal point 5 digits to the right. Then: 10^5 → When the decimal moved to the left, the exponent is positive. Then: $943{,}000 = 9.43 \times 10^5$)

5) 0.00005 (The exponent is negative 5. Then, move the decimal point to the left five digits. When the decimal moved to the right, the exponent is negative. Then: $5 \times 10^{-5} = 0.00005$)

6) 0.00032 (The exponent is negative 4. When the decimal moved to the right, the exponent is negative. Then: $3.2 \times 10^{-4} = 0.00032$)

7) 0.0000034 (The exponent is negative 6. When the decimal moved to the right, the exponent is negative. Then: $3.4 \times 10^{-6} = 0.0000034$)

8) 0.00000068 (The exponent is negative 7. Then, move the decimal point to the left five digits. When the decimal moved to the right, the exponent is negative. Then: $6.8 \times 10^{-7} = 0.00000068$)

Radicals

1) 54 (Find the values of $\sqrt{36}$ and $\sqrt{81}$. Then: $\sqrt{36} \times \sqrt{81} = 6 \times 9 = 54$)

2) 9 (Use this radical rule: $\sqrt[n]{x} \times \sqrt[n]{y} = \sqrt[n]{xy} \rightarrow \sqrt{3} \times \sqrt{27} = \sqrt{81}$. The square root of 81 is 9. Then: $\sqrt{3} \times \sqrt{27} = \sqrt{81} = 9$)

3) 27 (Use this radical rule: $\sqrt[n]{x} \times \sqrt[n]{y} = \sqrt[n]{xy} \rightarrow \sqrt{27} \times \sqrt{27} = \sqrt{27 \times 27} = \sqrt{(27)^2}$. Now use radical rule: $\sqrt[n]{a^n} = a$. Then: $\sqrt{27^2} = 27$)

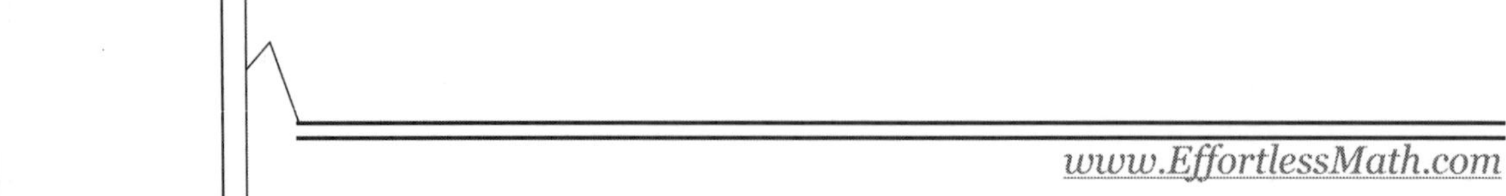

4) $2\sqrt{125}$ (Since we have the same values under the radical, we can add these two radicals: $\sqrt{125}+\sqrt{125}=2\sqrt{125}$)

5) $3\sqrt{7}$ (Since we have the same values under the radical, we can subtract these two radicals: $6\sqrt{7}-3\sqrt{7}=3\sqrt{7}$)

6) 80 (Use this radical rule: $\sqrt[n]{x}\times\sqrt[n]{y}=\sqrt[n]{xy}\rightarrow\sqrt{10}\times\sqrt{10}=\sqrt{10\times10}=\sqrt{10^2}$. Now use radical rule: $\sqrt[n]{a^n}=a\rightarrow\sqrt{10^2}=10$. Then: $4\sqrt{10}\times2\sqrt{10}=4\times2\times10=80$)

7) 42 (Use this radical rule: $\sqrt[n]{x}\times\sqrt[n]{y}=\sqrt[n]{xy}\rightarrow\sqrt{2}\times\sqrt{2}=\sqrt{2\times2}=\sqrt{2^2}$. Now use radical rule: $\sqrt[n]{a^n}=a\rightarrow\sqrt{2^2}=2$. Then: $7\sqrt{2}\times3\sqrt{2}=7\times3\times2=42$)

8) $5\sqrt{5}$ (Factor the number: $20=4\times5=2^2\times5$. Then: $\sqrt{20}=\sqrt{2^2\times5}=2\sqrt{5}$. Now we have the same values under the radical and we can subtract these two radicals: $7\sqrt{5}-\sqrt{20}=7\sqrt{5}-2\sqrt{5}=5\sqrt{5}$)

CHAPTER

7 Expressions and Variables

Math topics that you'll learn in this chapter:

- ☑ Simplifying Variable Expressions
- ☑ Simplifying Polynomial Expressions
- ☑ The Distributive Property
- ☑ Evaluating One Variable
- ☑ Evaluating Two Variables

Simplifying Variable Expressions

- In algebra, a variable is a letter used to stand for a number. The most common letters are x, y, z, a, b, c, m, and n.
- An algebraic expression is an expression that contains integers, variables, and math operations such as addition, subtraction, multiplication, division, etc.
- In an expression, we can combine "like" terms. (values with same variable and same power)

Examples:

Example 1. Simplify. $(4x + 2x + 4) =$
Solution: In this expression, there are three terms: $4x, 2x$, and 4. Two terms are "like terms": $4x$ and $2x$. Combine like terms. $4x + 2x = 6x$. Then: $(4x + 2x + 4) = 6x + 4$ (***remember you cannot combine variables and numbers.***)

Example 2. Simplify. $-2x^2 - 5x + 4x^2 - 9 =$
Solution: Combine "like" terms: $-2x^2 + 4x^2 = 2x^2$.
Then: $-2x^2 - 5x + 4x^2 - 9 = 2x^2 - 5x - 9$.

Practices:

Simplify each expression.

1) $(-7x + 3x + 5 + 18) =$
2) $(12x - 8x + 31) =$
3) $-4x + 3 - 4x =$
4) $2x^3 + 9x^3 - 9 =$
5) $-13 - 2x^2 + 2 =$
6) $10x + (28x - 17) =$
7) $(5x^2 - 15) - 9x^2 =$
8) $6x^2 + 78x - 8x =$

Simplifying Polynomial Expressions

- In mathematics, a polynomial is an expression consisting of variables and coefficients that involves only the operations of addition, subtraction, multiplication, and non–negative integer exponents of variables.

$$P(x) = a_n x^n + a_{n-1} x^{n-1} + \ldots + a_2 x^2 + a_1 x + a_0$$

- Polynomials must always be simplified as much as possible. It means you must add together any like terms. (values with same variable and same power)

Examples:

Example 1. Simplify this Polynomial Expressions. $3x^2 - 6x^3 - 2x^3 + 4x^4$
Solution: Combine "like" terms: $-6x^3 - 2x^3 = -8x^3$
Then: $3x^2 - 6x^3 - 2x^3 + 4x^4 = 3x^2 - 8x^3 + 4x^4$
Now, write the expression in standard form: $3x^2 - 8x^3 + 4x^4 = 4x^4 - 8x^3 + 3x^2$

Example 2. Simplify this expression. $(-5x^2 + 2x^3) - (3x^3 - 6x^2) =$
Solution: First, multiply $(-)$ into $(3x^3 - 6x^2)$:
$(-5x^2 + 2x^3) - (3x^3 - 6x^2) = -5x^2 + 2x^3 - 3x^3 + 6x^2$
Then combine "like" terms: $-5x^2 + 2x^3 - 3x^3 + 6x^2 = x^2 - x^3$
And write in standard form: $x^2 - x^3 = -x^3 + x^2$

Practices:

Simplify each polynomial.

1) $(5x^3 + x^2) - (11x + 3x^2) =$ ____________

2) $(4x^4 + 3x^3) - (x^3 + 5x^2) =$ ____________

3) $(9x^4 + 7x^2) - (2x^2 - 2x^4) =$ ____________

4) $12x - 10x^2 - 3(3x^2 + 2x^3) =$ ____________

5) $(10x^3 - 10) + 4(8x^2 - 5x^3) =$ ____________

Find more at bit.ly/2WT5gtn

The Distributive Property

- The distributive property (or the distributive property of multiplication over addition and subtraction) simplifies and solves expressions in the form of: $a(b+c)$ or $a(b-c)$
- The distributive property is multiplying a term outside the parentheses by the terms inside.
- Distributive Property rule: $a(b+c)=ab+ac$

Examples:

Example 1. Simply using the distributive property. $(-2)(x+3)$

Solution: Use Distributive Property rule: $a(b+c)=ab+ac$

$(-2)(x+3)=(-2\times x)+(-2)\times(3)=-2x-6$

Example 2. Simply. $(-5)(-2x-6)$

Solution: Use Distributive Property rule: $a(b+c)=ab+ac$

$(-5)(-2x-6)=(-5\times -2x)+(-5)\times(-6)=10x+30$

Practices:

Use the distributive property to simply each expression.

1) $4(4-2x)=$
2) $5(2+3x)=$
3) $(-4)(2x-6)=$
4) $(10x-2)(-5)=$
5) $(-11)(x-3)=$
6) $(3+2x)8=$
7) $12(7-2x)=$
8) $-(-9-13x)=$

Evaluating One Variable

- To evaluate one variable expressions, find the variable and substitute a number for that variable.
- Perform the arithmetic operations.

Examples:

Example 1. Calculate this expression for $x = 2$. $8 + 2x$

Solution: First, substitute 2 for x.
Then: $8 + 2x = 8 + 2(2)$
Now, use order of operation to find the answer: $8 + 2(2) = 8 + 4 = 12$

Example 2. Evaluate this expression for $x = -1$. $4x - 8$

Solution: First, substitute -1 for x.
Then: $4x - 8 = 4(-1) - 8$
Now, use order of operation to find the answer: $4(-1) - 8 = -4 - 8 = -12$

Practices:

Evaluate each expression using the value given.

1) $6 + x, x = 3$
2) $x - 11, x = 13$
3) $4x + 3, x = 2$
4) $5x - 13, x = -1$
5) $7 - x, x = 6$
6) $x + 5, x = 10$
7) $14x + 2, x = -3$
8) $x + (-9), x = -5$

Find more at bit.ly/3ppujQZ

Evaluating Two Variables

- To evaluate an algebraic expression, substitute a number for each variable.
- Perform the arithmetic operations to find the value of the expression.

Examples:

Example 1. Calculate this expression for $a = 2$ and $b = -1$. $(4a - 3b)$

Solution: First, substitute 2 for a, and -1 for b.
Then: $4a - 3b = 4(2) - 3(-1)$
Now, use order of operation to find the answer: $4(2) - 3(-1) = 8 + 3 = 11$

Example 2. Evaluate this expression for $x = -2$ and $y = 2$. $(3x + 6y)$

Solution: Substitute -2 for x, and 2 for y.
Then: $3x + 6y = 3(-2) + 6(2) = -6 + 12 = 6$

Practices:

✍ ***Evaluate each expression using the values given.***

1) $5x + 3y$,
   $x = 2, y = 4$
2) $2x + 11y$,
   $x = 8, y = 1$
3) $8a + 7b$,
   $a = 1, b = 3$
4) $4x - y + 6$,
   $x = 4, y = 9$
5) $2a + 24 - 3b$,
   $a = -2, b = 2$
6) $3(6x - 2y)$,
   $x = 6, y = 8$
7) $14a + 4b$,
   $a = 2, b = 2$
8) $8x \div 2y$,
   $x = 3, y = 2$

Chapter 7: Answers

Simplifying Variable Expressions

1) $-4x+23$ (In this expression, there are four terms: $-7x, 3x, 5$ and 18. Two terms are "like terms": $-7x$ and $3x$. also 5 and18 are "like terms". Combine like terms. $-7x+3x=-4x$ and $5+18=23$. Then:

   $(-7x+3x+5+18)=-4x+23$.

2) $4x+31$ (Combine "like" terms: $12x-8x=4x$. Then: $(12x-8x+31)=4x+31$.

3) $-8x+3$ (Combine "like" terms: $-4x-4x=-8x$. Then: $-4x+3-4x=-8x+3$)

4) $11x^3-9$ (Combine "like" terms: $2x^3+9x^3=11x^3$. Then: $2x^3+9x^3-9=$

   $11x^3-9$)

5) $-2x^2-11$ (Combine "like" terms: $-13+2=-11$. Then: $-13-2x^2+2=-11-2x^2$. Write in standard form (biggest powers first): $-11-2x^2=-2x^2-11$)

6) $38x-17$ (Combine "like" terms: $10x+28x=38x$. Then:

   $10x+(28x-17)=38x-17$)

7) $-4x^2-15$ (Combine "like" terms: $5x^2-9x^2=-4x^2$. Then:

   $(5x^2-15)-9x^2=-4x^2-15$)

8) $6x^2+70x$ (Combine "like" terms:$78x-8x=70x$. Then:

   $6x^2+78x-8x=6x^2+70x$)

Simplifying Polynomial Expressions

1) $5x^3-2x^2-11x$ (First, multiply $(-)$ into $(11x+3x^2)$: $(5x^3+x^2)-(11x+3x^2)=5x^3+x^2-11x-3x^2$. Then combine "like" terms: $5x^3+x^2-11x-3x^2=5x^3-2x^2-11x$)

2) $4x^4 + 2x^3 - 5x^2$ (First, multiply $(-)$ into $(x^3 + 5x^2)$: $(4x^4 + 3x^3) - (x^3 + 5x^2) = 4x^4 + 3x^3 - x^3 - 5x^2$. Then combine "like" terms: $4x^4 + 3x^3 - x^3 - 5x^2 = 4x^4 + 2x^3 - 5x^2$.)

3) $11x^4 + 5x^2$ (First, multiply $(-)$ into $(2x^2 - 2x^4)$: $(9x^4 + 7x^2) - (2x^2 - 2x^4) = 9x^4 + 7x^2 - 2x^2 + 2x^4$. Then combine "like" terms: $9x^4 + 7x^2 - 2x^2 + 2x^4 = 11x^4 + 5x^2$.)

4) $-6x^3 - 19x^2 + 12x$ (First, multiply (-3) into $(3x^2 + 2x^3)$: $12x - 10x^2 - 3(3x^2 + 2x^3) = 12x - 10x^2 - 9x^2 - 6x^3$. Then combine "like" terms: $12x - 10x^2 - 9x^2 - 6x^3 = 12x - 19x^2 - 6x^3$. And write in standard form: $12x - 19x^2 - 6x^3 = -6x^3 - 19x^2 + 12x$)

5) $-10x^3 + 32x^2 - 10$ (First, multiply (4) into $(8x^2 - 5x^3)$: $(10x^3 - 10) + 4(8x^2 - 5x^3) = 10x^3 - 10 + 32x^2 - 20x^3$. Then combine "like" terms: $10x^3 - 10 + 32x^2 - 20x^3 = -10x^3 - 10 + 32x^2$. And write in standard form: $-10x^3 - 10 + 32x^2 = -10x^3 + 32x^2 - 10$)

The Distributive Property

1) $-8x + 16$ (Use Distributive Property rule: $a(b + c) = ab + ac \rightarrow 4(4 - 2x) = (4 \times 4) + (4) \times (-2x) = 16 - 8x$. And write in standard form: $16 - 8x = -8x + 16$)

2) $15x + 10$ (Use Distributive Property rule: $a(b + c) = ab + ac \rightarrow 5(2 + 3x) = (5 \times 2) + (5) \times (3x) = 10 + 15x$. And write in standard form: $10 + 15x = 15x + 10$)

3) $-8x + 24$ (Use Distributive Property rule: $a(b + c) = ab + ac \rightarrow (-4)(2x - 6) = (-4) \times (2x) + (-4) \times (-6) = -8x + 24$.)

4) $-50x + 10$ (Use Distributive Property rule: $a(b + c) = ab + ac \rightarrow (10x - 2)(-5) = (10x) \times (-5) + (-2) \times (-5) = -50x + 10$.)

5) $-11x + 33$ (Use Distributive Property rule: $a(b + c) = ab + ac \rightarrow (-11)(x - 3) = (-11) \times (x) + (-11) \times (-3) = -11x + 33$.)

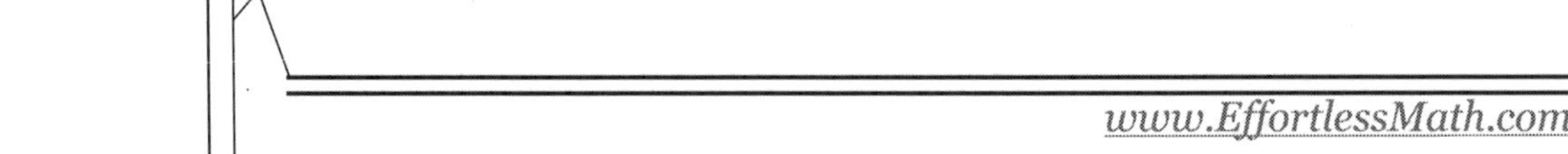

6) $16x + 24$ (Use Distributive Property rule: $a(b + c) = ab + ac \rightarrow (3 + 2x)8 = (3 \times 8) + (2x) \times (8) = 24 + 16x$. And write in standard form:$24 + 16x = 16x + 24$)

7) $-24x + 84$ (Use Distributive Property rule: $a(b + c) = ab + ac \rightarrow 12(7 - 2x) = (12 \times 7) + (12) \times (-2x) = 84 - 24x$. And write in standard form: $84 + 24x = -24x + 84$)

8) $13x + 9$ (Use Distributive Property rule: $a(b + c) = ab + ac \rightarrow -(-9 - 13x) = (-1) \times (-9) + (-1) \times (13x) = 9 - 13x$. And write in standard form: $9 - 13x = 13x + 9$)

Evaluating One Variable

1) 9 (First, substitute 3 for x. Then: $6 + x\,, x = 3 \rightarrow 6 + (3)$. Now, use order of operation to find the answer:$6 + (3) = 6 + 3 = 9$)

2) 2 (First, substitute 13 for x. Then: $x - 11, x = 13 \rightarrow (13) - 11$. Now, use order of operation to find the answer:$(13) - 11 = 13 - 11 = 2$)

3) 11 (First, substitute 2 for x. Then: $4x + 3, x = 2 \rightarrow 4(2) + 3$. Now, use order of operation to find the answer: $4(2) + 3 = 8 + 3 = 11$)

4) -18 (First, substitute -1 for x. Then: $5x - 13, x = -1 \rightarrow 5(-1) - 13$. Now, use order of operation to find the answer: $5(-1) - 13 = -5 - 13 = -18$)

5) 1 (First, substitute 6 for x. Then: $7 - x\,, x = 6 \rightarrow 7 - (6)$. Now, use order of operation to find the answer: $7 - (6) = 7 - 6 = 1$)

6) 15 (First, substitute 10 for x. Then: $x + 5, x = 10 \rightarrow (10) + 5$. Now, use order of operation to find the answer: $(10) + 5 = 10 + 5 = 15$)

7) -40 (First, substitute -3 for x. Then:$14x + 2, x = -3 \rightarrow 14(-3) + 2$. Now, use order of operation to find the answer: $14(-3) + 2 = -42 + 2 = -40$)

8) -14 (First, substitute -5 for x. Then: $x + (-9), x = -5 \rightarrow (-5) + (-9)$. Now, use order of operation to find the answer: $(-5) + (-9) = -5 - 9 = -14$)

Evaluating Two Variables

1) 22 (First, substitute 2 for x, and 4 for y. then: $5x + 3y = 5(2) + 3(4)$. Now, use order of operation to find the answer:$5(2) + 3(4) = 10 + 12 = 22$)
2) 27 (First, substitute 8 for x, and 1 for y. then: $2x + 11y = 2(8) + 11(1)$. Now, use order of operation to find the answer: $2(8) + 11(1) = 16 + 11 = 27$)
3) 29 (First, substitute 1 for a, and 3 for b. then: $8a + 7b = 8(1) + 7(3)$. Now, use order of operation to find the answer: $8(1) + 7(3) = 8 + 21 = 29$.)
4) 13 (First, substitute 4 for x, and 9 for y. then: $4x - y + 6 = 4(4) - (9) + 6$. Now, use order of operation to find the answer: $4(4) - (9) + 6 = 16 - 9 + 6 = 13$)
5) 14 (First, substitute -2 for a, and 2 for b. then $2a + 24 - 3b = 2(-2) + 24 - 3(2)$. Now, use order of operation to find the answer: $2(-2) + 24 - 3(2) = -4 + 24 - 6 = 14$)
6) 60 (First, substitute 6 for x, and 8 for y. then $3(6x - 2y) = 3(6 \times (6) - 2(8))$. Now, use order of operation to find the answer: $3(6 \times (6) - 2(8)) = 3(36 - 16) = 3(20) = 60$)
7) 36 (First, substitute 2 for a, and 2 for b. then $14a + 4b = 14(2) + 4(2)$. Now, use order of operation to find the answer: $14(2) + 4(2) = 28 + 8 = 36$)
8) 6 (First, substitute 3 for x, and 2 for y. then $8x \div 2y = 8(3) \div 2(2)$. Now, use order of operation to find the answer: $8(3) \div 2(2) = 24 \div 4 = 6$)

CHAPTER

8 Equations and Inequalities

Math topics that you'll learn in this chapter:

☑ One-Step Equations

☑ Multi-Step Equations

☑ System of Equations

☑ Graphing Single–Variable Inequalities

☑ One-Step Inequalities

☑ Multi-Step Inequalities

One–Step Equations

- The values of two expressions on both sides of an equation are equal. Example: $ax = b$. In this equation, ax is equal to b.
- Solving an equation means finding the value of the variable.
- You only need to perform one Math operation to solve the one-step equations.
- To solve a one-step equation, find the inverse (opposite) operation is being performed.
- The inverse operations are:
  - ❖ Addition and subtraction
  - ❖ Multiplication and division

Examples:

Example 1. Solve this equation for x. $4x = 16 \rightarrow x = ?$

Solution: Here, the operation is multiplication (variable x is multiplied by 4) and its inverse operation is division. To solve this equation, divide both sides of the equation by 4: $4x = 16 \rightarrow \frac{4x}{4} = \frac{16}{4} \rightarrow x = 4$

Example 2. Solve this equation. $x + 8 = 0 \rightarrow x = ?$

Solution: In this equation, 8 is added to the variable x. The inverse operation of addition is subtraction. To solve this equation, subtract 8 from both sides of the equation: $x + 8 - 8 = 0 - 8$. Then: $x + 8 - 8 = 0 - 8 \rightarrow x = -8$

Practices:

Solve each equation.

1) $26 = -8 + x, x =$ ____
2) $x - 12 = -38, x =$ ____
3) $x + 15 = -11, x =$ ____
4) $10 = x - 27, x =$ ____
5) $4 + x = -21, x =$ ____
6) $x - 7 = -33\ x =$ ____

Multi–Step Equations

- To solve a multi-step equation, combine "like" terms on one side.
- Bring variables to one side by adding or subtracting.
- Simplify using the inverse of addition or subtraction.
- Simplify further by using the inverse of multiplication or division.
- Check your solution by plugging the value of the variable into the original equation.

Examples:

Example 1. Solve this equation for x. $4x + 8 = 20 - 2x$

Solution: First, bring variables to one side by adding $2x$ to both sides. Then:

$4x + 8 + 2x = 20 - 2x + 2x \rightarrow 4x + 8 + 2x = 20$.

Simplify: $6x + 8 = 20$. Now, subtract 8 from both sides of the equation:

$6x + 8 - 8 = 20 - 8 \rightarrow 6x = 12 \rightarrow$ Divide both sides by 6:

$6x = 12 \rightarrow \frac{6x}{6} = \frac{12}{6} \rightarrow x = 2$

Let's check this solution by substituting the value of 2 for x in the original equation: $x = 2 \rightarrow 4x + 8 = 20 - 2x \rightarrow 4(2) + 8 = 20 - 2(2) \rightarrow 16 = 16$

The answer $x = 2$ is correct.

Example 2. Solve this equation for x. $-5x + 4 = 24$

Solution: Subtract 4 from both sides of the equation.

$-5x + 4 = 24 \rightarrow -5x + 4 - 4 = 24 - 4 \rightarrow -5x = 20$

Divide both sides by -5, then: $-5x = 20 \rightarrow \frac{-5x}{-5} = \frac{20}{-5} \rightarrow x = -4$

Now, check the solution: $x = -4 \rightarrow -5x + 4 = 24 \rightarrow -5(-4) + 4 = 24 \rightarrow 24 = 24$

The answer $x = -4$ is correct.

Practices:

Solve each equation.

1) $-4(x + 5) = 16$
2) $3(4 + x) = 27$
3) $-36 + 2x = 14x$
4) $5x + 27 = -2x - 22$
5) $15 - 4x = -9 - 3x$
6) $19 - 6x = 10 + 3x$

Find more at bit.ly/3nQbSEB

System of Equations

- A system of equations contains two equations and two variables. For example, consider the system of equations: $x - y = 1$ and $x + y = 5$
- The easiest way to solve a system of equations is using the elimination method. The elimination method uses the addition property of equality. You can add the same value to each side of an equation.
- For the first equation above, you can add $x + y$ to the left side and 5 to the right side of the first equation: $x - y + (x + y) = 1 + 5$. Now, if you simplify, you get: $x - y + (x + y) = 1 + 5 \rightarrow 2x = 6 \rightarrow x = 3$. Now, substitute 3 for the x in the first equation: $3 - y = 1$. By solving this equation, $y = 2$

Example:

What is the value of $x + y$ in this system of equations?

$$\begin{cases} 2x + 4y = 12 \\ 4x - 2y = -16 \end{cases}$$

Solution: Solving a System of Equations by Elimination:

Multiply the first equation by (-2), then add it to the second equation.

$$\frac{\begin{matrix} -2(2x + 4y = 12) \\ 4x - 2y = -16 \end{matrix}}{} \Rightarrow \begin{matrix} -4x - 8y = -24 \\ 4x - 2y = -16 \end{matrix} \Rightarrow (-4x) + 4x - 8y - 2y = -24 - 16 \Rightarrow$$

$-10y = -40 \Rightarrow y = 4$

Plug in the value of y into one of the equations and solve for x.

$2x + 4(4) = 12 \Rightarrow 2x + 16 = 12 \Rightarrow 2x = -4 \Rightarrow x = -2$

Thus, $x + y = -2 + 4 = 2$

Practices:

✍ ***Solve each system of equations.***

1) $3x - y = 7$, $2x + 3y = 1$ $x =$ ____ $y =$ ____

2) $x + y = 6$, $-3x + y = 2$ $x =$ ____ $y =$ ____

3) $2x + 4y = -10$, $6x + 3y = 6$ $x =$ ____ $y =$ ____

4) $2x + 2y = 26$, $7x + 2y = 31$ $x =$ ____ $y =$ ____

Graphing Single–Variable Inequalities

- An inequality compares two expressions using an inequality sign.
- Inequality signs are: "less than" <, "greater than" >, "less than or equal to" ≤, and "greater than or equal to" ≥.
- To graph a single–variable inequality, find the value of the inequality on the number line.
- For less than (<) or greater than (>) draw an open circle on the value of the variable. If there is an equal sign too, then use a filled circle.
- Draw an arrow to the right for greater or to the left for less than.

Examples:

Example 1. Draw a graph for this inequality. $x > 2$

Solution: Since the variable is greater than 2, then we need to find 2 in the number line and draw an open circle on it. Then, draw an arrow to the right.

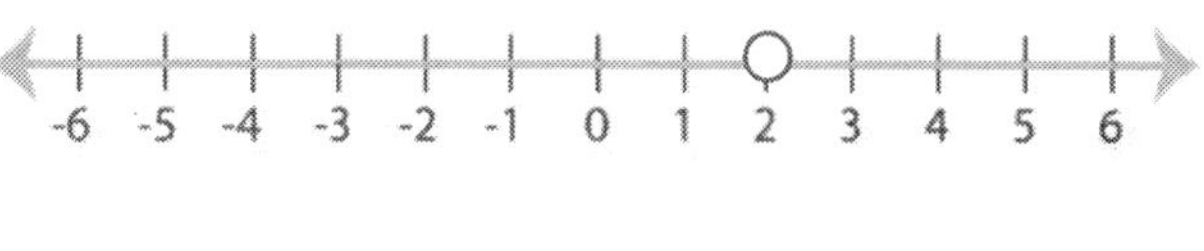

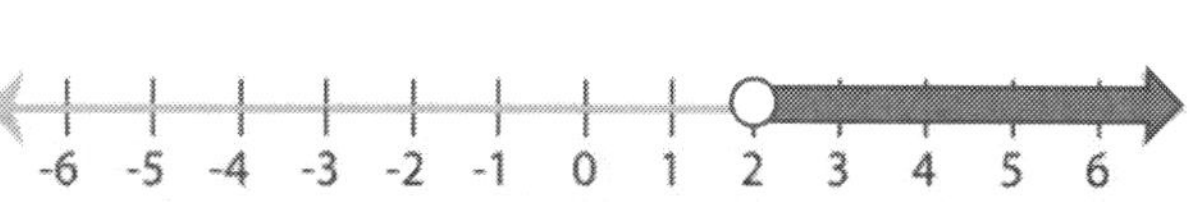

Example 2. Graph this inequality. $x \leq -3$.

Solution: Since the variable is less than or equal to -3, then we need to find -3 on the number line and draw a filled circle on it. Then, draw an arrow to the left.

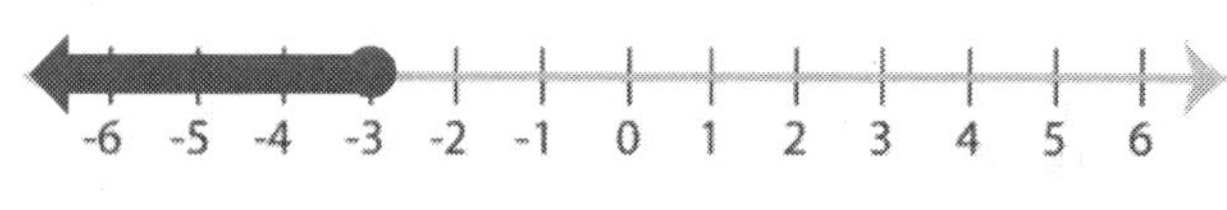

Practices:

Draw a graph for each inequality.

1) $x > 4$

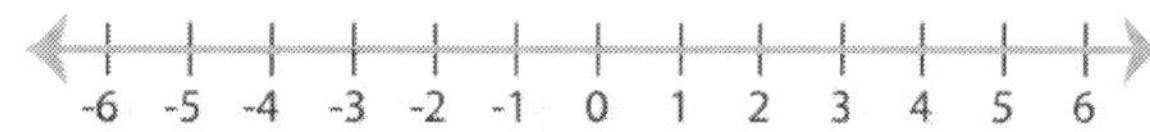

2) $x < 5$

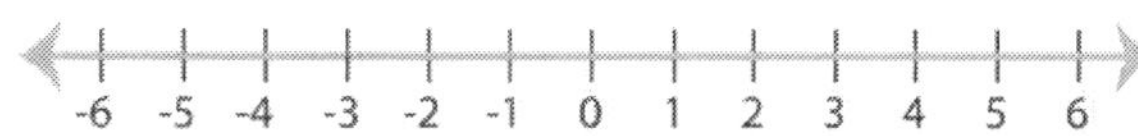

3) $x \leq 0$

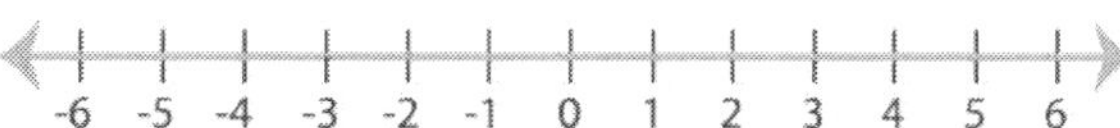

4) $x > -2$

-6 -5 -4 -3 -2 -1 0 1 2 3 4 5 6

One–Step Inequalities

- An inequality compares two expressions using an inequality sign.
- Inequality signs are: "less than" $<$, "greater than" $>$, "less than or equal to" $\leq$, and "greater than or equal to" $\geq$.
- You only need to perform one Math operation to solve the one-step inequalities.
- To solve one-step inequalities, find the inverse (opposite) operation is being performed.
- For dividing or multiplying both sides by negative numbers, flip the direction of the inequality sign.

Examples:

Example 1. Solve this inequality for x. $x + 5 \geq 4$

Solution: The inverse (opposite) operation of addition is subtraction. In this inequality, 5 is added to x. To isolate x we need to subtract 5 from both sides of the inequality.

Then: $x + 5 \geq 4 \rightarrow x + 5 - 5 \geq 4 - 5 \rightarrow x \geq -1$. The solution is: $x \geq -1$

Example 2. Solve. $-3x \leq 6$

Solution: -3 is multiplied to x. Divide both sides by -3. Remember when dividing or multiplying both sides of an inequality by negative numbers, flip the direction of the inequality sign.

Then: $-3x \leq 6 \rightarrow \frac{-3x}{-3} \geq \frac{6}{-3} \rightarrow x \geq -2$

Practices:

Solve each inequality and graph it.

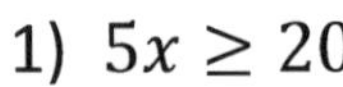

1) $5x \geq 20$ -6 -5 -4 -3 -2 -1 0 1 2 3 4 5 6

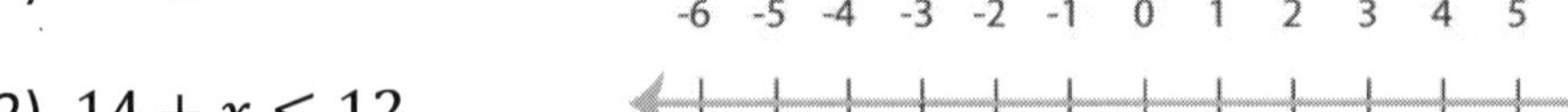

2) $14 + x \leq 12$ -6 -5 -4 -3 -2 -1 0 1 2 3 4 5 6

3) $x - 9 \leq -4$ -6 -5 -4 -3 -2 -1 0 1 2 3 4 5 6

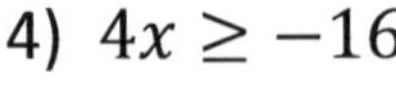

4) $4x \geq -16$ -6 -5 -4 -3 -2 -1 0 1 2 3 4 5 6

Multi–Step Inequalities

- To solve a multi-step inequality, combine "like" terms on one side.
- Bring variables to one side by adding or subtracting.
- Isolate the variable.
- Simplify using the inverse of addition or subtraction.
- Simplify further by using the inverse of multiplication or division.
- For dividing or multiplying both sides by negative numbers, flip the direction of the inequality sign.

Examples:

Example 1. Solve this inequality. $8x - 2 \leq 14$
Solution: In this inequality, 2 is subtracted from $8x$. The inverse of subtraction is addition. Add 2 to both sides of the inequality:
$8x - 2 + 2 \leq 14 + 2 \rightarrow 8x \leq 16$
Now, divide both sides by 8. Then: $8x \leq 16 \rightarrow \frac{8x}{8} \leq \frac{16}{8} \rightarrow x \leq 2$
The solution to this inequality is $x \leq 2$.

Example 2. Solve this inequality. $3x + 9 < 12$
Solution: First, subtract 9 from both sides: $3x + 9 - 9 < 12 - 9$
Then simplify: $3x + 9 - 9 < 12 - 9 \rightarrow 3x < 3$
Now divide both sides by 3: $\frac{3x}{3} < \frac{3}{3} \rightarrow x < 1$

Practices:

Solve each inequality.

1) $4x - 3 \leq 5$
2) $7x + 2 \leq 9$
3) $3 + 5x > 13$
4) $3(x + 2) \leq 18$
5) $2x - 15 \geq 7$
6) $5x - 21 < 4$

Find more at bit.ly/2WK1xOr

Chapter 8: Answers

One–Step Equations

1) $x = 34$ (Here, the operation is subtraction and its inverse operation is addition. To solve this equation, add 8 to both sides of the equation: $26 + 8 = -8 + x - 8 \rightarrow x = 34$)

2) $x = -26$ (Here, the operation is subtraction and its inverse operation is addition. To solve this equation, add 12 to both sides of the equation: $x - 12 + 12 = -38 + 12 \rightarrow x = -26$)

3) $x = -26$ (In this equation, 15 is added to the variable x. The inverse operation of addition is subtraction. To solve this equation, subtract 15 from both sides of the equation: $x + 15 - 15 = -11 - 15$. Then: $x = -26$)

4) $x = 37$ (Here, the operation is subtraction and its inverse operation is addition. To solve this equation, add 27 to both sides of the equation: $10 + 27 = x - 27 + 27 \rightarrow x = 37$)

5) $x = -25$ (In this equation, 4 is added to the variable x. The inverse operation of addition is subtraction. To solve this equation, subtract 4 from both sides of the equation: $4 + x - 4 = -21 - 4$. Then: $x = -25$)

6) $x = -26$ (Here, the operation is subtraction and its inverse operation is addition. To solve this equation, add 7 to both sides of the equation: $x - 7 + 7 = -33 + 7 \rightarrow x = -26$)

Multi–Step Equations

1) $x = 1$ (First use Distributive Property: $-4(x + 5) = -4x + 20$. Now, subtract 20 from both sides of the equation: $-4x + 20 = 16 \rightarrow -4x + 20 - 20 = 16 - 20$. Now simplify: $-4x = -4 \rightarrow$ Divide both sides by -4: $-4x = -4 \rightarrow \frac{-4x}{-4} = \frac{-4}{-4} \rightarrow x = 1$)

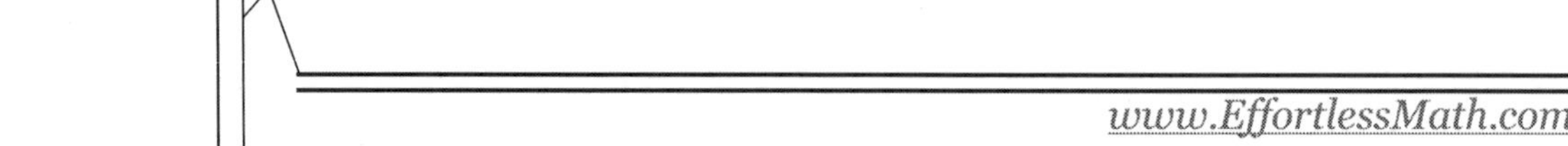

2) $x = 5$ (First use Distributive Property: $3(4 + x) = 12 + 3x$. Now, subtract 12 from both sides of the equation: $12 + 3x = 27 \rightarrow 12 + 3x - 12 = 27 - 12$. Now simplify: $3x = 15 \rightarrow$ Divide both sides by 3: $3x = 15 \rightarrow \frac{3x}{3} = \frac{15}{3} \rightarrow x = 5$)

3) $x = -3$ (First, bring variables to one side by subtracting $2x$ to both sides. Then: $-36 + 2x - 2x = 14x - 2x \rightarrow -36 = 12x$. Now, divide both sides by 12: $-36 = 12x \rightarrow \frac{-36}{12} = \frac{12x}{12} \rightarrow x = -3$)

4) $x = -7$ (First, bring variables to one side by adding $2x$ to both sides. Then: $5x + 27 + 2x = -2x - 22 + 2x \rightarrow 7x + 27 = -22$. Now, subtract 27 from both sides of the equation: $7x + 27 - 27 = -22 - 27 \rightarrow 7x = -49 \rightarrow$Divide both sides by 7: $7x = -49 \rightarrow \frac{7x}{7} = \frac{-49}{7} \rightarrow x = -7$)

5) $x = 24$ (Bring variables to one side by subtracting $3x$ to both sides. Then: $15 - 4x + 3x = -9 - 3x + 3x \rightarrow -x + 15 = -9$. Now, subtract 15 from both sides of the equation: $-x + 15 - 15 = -9 - 15 \rightarrow -x = -24 \rightarrow$Divide both sides by -1: $-x = -24 \rightarrow \frac{-x}{-1} = \frac{-24}{-1} \rightarrow x = 24$)

6) $x = 1$ (Bring variables to one side by adding $3x$ to both sides. Then: $19 - 6x - 3x = 10 + 3x - 3x \rightarrow 19 - 9x = 10$. Now, subtract 19 from both sides of the equation: $19 - 9x - 19 = 10 - 19 \rightarrow -9x = -9 \rightarrow$Divide both sides by -9: $-9x = -9 \rightarrow \frac{-9x}{-9} = \frac{-9}{-9} \rightarrow x = 1$)

System of Equations

1) $x = 2, y = -1$ (Multiply the first equation by 3, then add it to the second equation.

$$\begin{array}{c} 3(3x - y = 7) \\ \underline{2x + 3y = 1} \end{array} \Rightarrow \begin{array}{c} 9x - 3y = 21 \\ 2x + 3y = 1 \end{array} \Rightarrow 9x + 2x - 3y + 3y = 21 + 1 \Rightarrow 11x = 22 \Rightarrow$$

$x = 2$

Plug in the value of x into one of the equations and solve for y.

$3(2) - y = 7 \Rightarrow 6 - y = 7 \Rightarrow -y = 1 \Rightarrow y = -1$)

2) $x = 1, y = 5$ (Multiply the first equation by 3, then add it to the second equation.

$$\begin{array}{l} 3(x + y = 6) \\ \underline{-3x + y = 2} \end{array} \Rightarrow \begin{array}{l} 3x + 3y = 18 \\ -3x + y = 2 \end{array} \Rightarrow 3x - 3x + 3y + y = 18 + 2 \Rightarrow 4y = 20 \Rightarrow$$

$$y = 5$$

Plug in the value of y into one of the equations and solve for x.

$x + 5 = 6 \Rightarrow x = 1$)

3) $x = 3, y = -4$ (Multiply the first equation by (-3), then add it to the second equation.

$$\begin{array}{c} -3(2x + 4y = -10) \\ \underline{6x + 3y = 6} \end{array} \Rightarrow \begin{array}{c} -6x - 12y = 30 \\ 6x + 3y = 6 \end{array} \Rightarrow -6x + 6x - 12y + 3y = 30 + 6 \Rightarrow$$

$$-9y = 36 \Rightarrow y = -4$$

Plug in the value of y into one of the equations and solve for x.

$2x + 4(-4) = -10 \Rightarrow 2x - 16 = -10 \Rightarrow 2x = -10 + 16 \Rightarrow 2x = 6 \Rightarrow x = 3$)

4) $x = 1, y = 12$ (Multiply the first equation by (-1), then add it to the second equation.

$$\begin{array}{c} -1(2x + 2y = 26) \\ \underline{7x + 2y = 31} \end{array} \Rightarrow \begin{array}{c} -2x - 2y = -26 \\ 7x + 2y = 31 \end{array} \Rightarrow -2x + 7x - 2y + 2y = -26 + 31 \Rightarrow$$

$$5x = 5 \Rightarrow x = 1$$

Plug in the value of x into one of the equations and solve for y.

$2(1) + 2y = 26 \Rightarrow 2 + 2y = 26 \Rightarrow 2y = 26 - 2 \Rightarrow 2y = 24 \Rightarrow y = 12$)

Graphing Single–Variable Inequalities

1) Since the variable is greater than 4, then we need to find 2 in the number line and draw an open circle on it. Then, draw an arrow to the right.

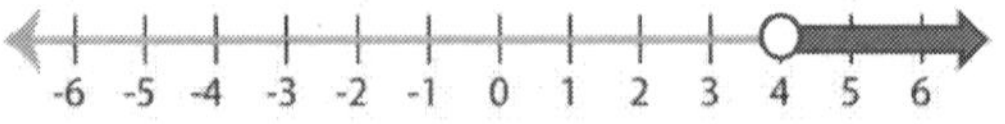

2) Since the variable is less than 5, then we need to find 5 on the number line and draw an open circle on it. Then, draw an arrow to the left.

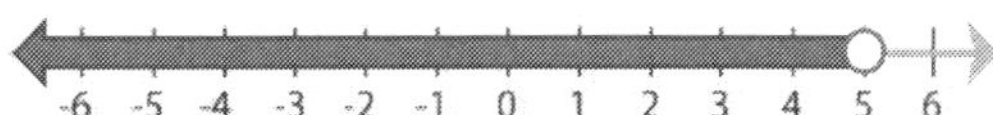

3) Since the variable is less than or equal to 0, then we need to find 0 on the number line and draw a filled circle on it. Then, draw an arrow to the left.

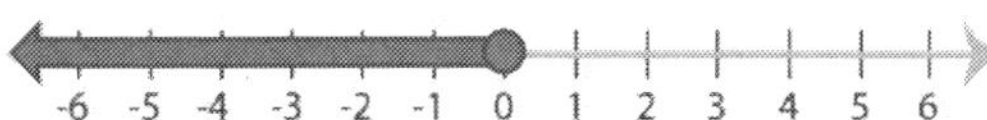

4) Since the variable is greater than -2, then we need to find -2 in the number line and draw an open circle on it. Then, draw an arrow to the right.

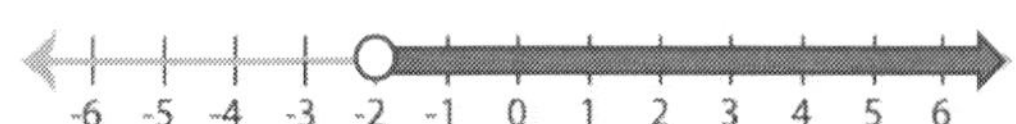

One–Step Inequalities

1) $x \geq 4$ (5 is multiplied to x. Divide both sides by 5. Then: $5x \geq 20 \rightarrow \frac{5x}{5} \geq \frac{20}{5} \rightarrow x \geq 4$)

2) $x \leq -2$ (The inverse (opposite) operation of addition is subtraction. In this inequality, 14 is added to x. To isolate x we need to subtract 14 from both sides of the inequality. Then: $14 + x \leq 12 \rightarrow 14 + x - 14 \leq 12 - 14 \rightarrow x \leq -2$. The solution is: $x \leq -2$)

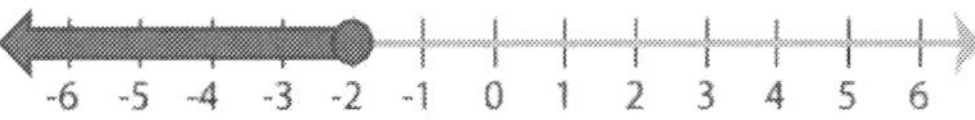

3) $x \leq 5$ (9 is subtracted from x. Add 9 to both sides. $x - 9 \leq -4 \rightarrow$ $x - 9 + 9 \leq -4 + 9 \rightarrow x \leq 5$)

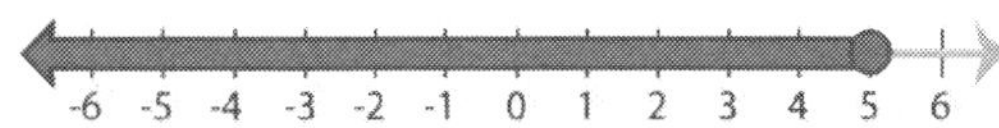

4) $x \geq -4$ (4 is multiplied to x. Divide both sides by 4. Then: $4x \geq -16 \rightarrow$ $\frac{4x}{4} \geq \frac{-16}{4} \rightarrow x \geq -4$)

Multi–Step Inequalities

1) (In this inequality, 3 is subtracted from $4x$. The inverse of subtraction is addition. Add 3 to both sides of the inequality: $4x - 3 + 3 \leq 5 + 3 \rightarrow 4x \leq 8$. Now, divide both sides by 4. Then: $4x \leq 8 \rightarrow \frac{4x}{4} \leq \frac{8}{4} \rightarrow x \leq 2$. The solution of this inequality is $x \leq 2$.)

2) (First, subtract 2 from both sides: $7x + 2 - 2 \leq 9 - 2$. Then simplify: $7x + 2 - 2 \leq 9 - 2 \rightarrow 7x \leq 7$. Now divide both sides by 7: $\frac{7x}{7} \leq \frac{7}{7} \rightarrow x \leq 1$)

3) (First, subtract 3 from both sides: $3 + 5x - 3 > 13 - 3$. Then simplify: $3 + 5x - 3 > 13 - 3 \rightarrow 5x > 10$. Now divide both sides by 5: $\frac{5x}{5} > \frac{10}{5} \rightarrow x > 2$)

4) (First, multiply 3 into $(x + 2)$: $3(x + 2) = 3x + 6$. second, subtract 6 from both sides: $3x + 6 - 6 \leq 18 - 6$. Then simplify: $3x + 6 - 6 \leq 18 - 6 \rightarrow 3x \leq 12$. Now divide both sides by 3: $\frac{3x}{3} \leq \frac{12}{3} \rightarrow x \leq 4$)

5) (First, add 15 to both sides: $2x - 15 + 15 \geq 7 + 15$. Then simplify: $2x - 15 + 15 \geq 7 + 15 \rightarrow 2x \geq 22$. Now divide both sides by 2: $\frac{2x}{2} \geq \frac{22}{2} \rightarrow x \geq 11$)

6) (First, add 21 to both sides: $5x - 21 + 21 < 4 + 21$. Then simplify: $5x - 21 + 21 < 4 + 21 \rightarrow 5x < 25$. Now divide both sides by 5: $\frac{5x}{5} < \frac{25}{5} \rightarrow x < 5$)

CHAPTER 9 Lines and Slope

Math topics that you'll learn in this chapter:

☑ Finding Slope

☑ Graphing Lines Using Slope–Intercept Form

☑ Writing Linear Equations

☑ Finding Midpoint

☑ Finding Distance of Two Points

☑ Graphing Linear Inequalities

Finding Slope

- The slope of a line represents the direction of a line on the coordinate plane.
- A coordinate plane contains two perpendicular number lines. The horizontal line is x and the vertical line is y. The point at which the two axes intersect is called the origin. An ordered pair (x, y) shows the location of a point.
- A line on a coordinate plane can be drawn by connecting two points.
- To find the slope of a line, we need the equation of the line or two points on the line.
- The slope of a line with two points A (x_1, y_1) and B (x_2, y_2) can be found by using this formula: $\frac{y_2-y_1}{x_2-x_1} = \frac{rise}{run}$
- The equation of a line is typically written as $y = mx + b$ where m is the slope and b is the y-intercept.

Examples:

Example 1. Find the slope of the line through these two points:
A$(1, -6)$ and $B(3, 2)$.

Solution: $Slop = \frac{y_2-y_1}{x_2-x_1}$. Let (x_1, y_1) be A$(1, -6)$ and (x_2, y_2) be $B(3, 2)$.
(Remember, you can choose any point for (x_1, y_1) and (x_2, y_2)).
Then: slope $= \frac{y_2-y_1}{x_2-x_1} = \frac{2-(-6)}{3-1} = \frac{8}{2} = 4$
The slope of the line through these two points is 4.

Example 2. Find the slope of the line with equation $y = -2x + 8$
Solution: When the equation of a line is written in the form of $y = mx + b$, the slope is m. In this line: $y = -2x + 8$, the slope is -2.

Practices:

✍ ***Find the slope of the line through each pair of points.***

1) $(3, 2), (4, 6)$
2) $(2, 1), (3, 4)$
3) $(0, 3), (1,0)$
4) $(4, 3), (8, 5)$
5) $(9, 8), (6, 23)$
6) $(10, 8), (12, 9)$
7) $(2, 7), (1, 8)$
8) $(-5, -2), (-3, 6)$

Graphing Lines Using Slope–Intercept Form

- Slope–intercept form of a line: given the slope m and the y–intercept (the intersection of the line and y-axis) b, then the equation of the line is:

$$y = mx + b$$

- To draw the graph of a linear equation in a slope-intercept form on the xy coordinate plane, find two points on the line by plugging two values for x and calculating the values of y.
- You can also use the slope (m) and one point to graph the line.

Example:

Sketch the graph of $y = 2x - 4$.

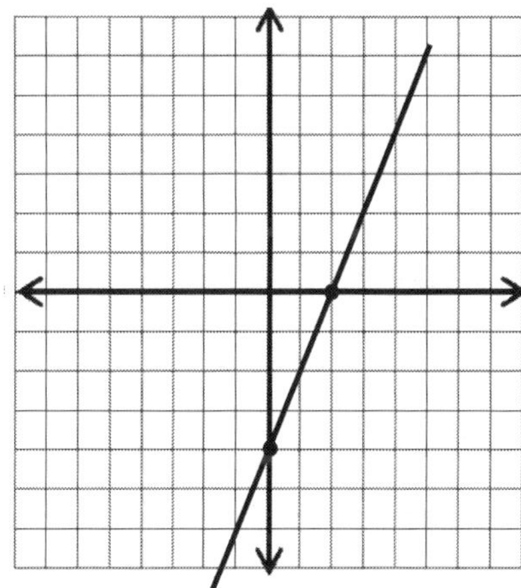

Solution: To graph this line, we need to find two points. When x is zero the value of y is -4. And when x is 2 the value of y is 0.

$$x = 0 \rightarrow y = 2(0) - 4 = -4,$$
$$y = 0 \rightarrow 0 = 2x - 4 \rightarrow x = 2$$

Now, we have two points:

$(0, -4)$ and $(2, 0)$.

Find the points on the coordinate plane and graph the line. Remember that the slope of the line is 2.

Practices:

Sketch the graph of each line.

1) $y = 3x + 3$

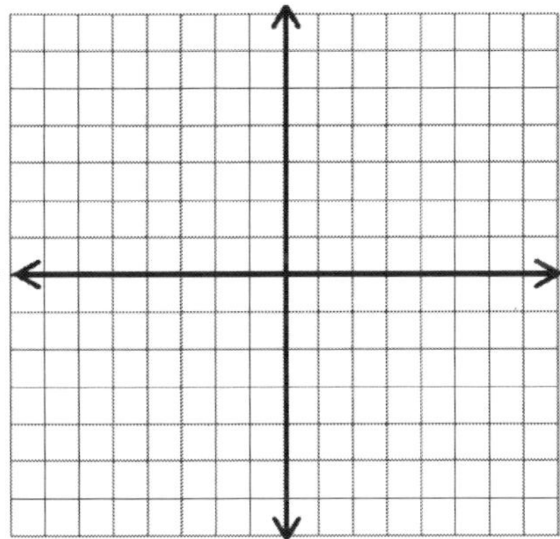

2) $y = 2x - 6$

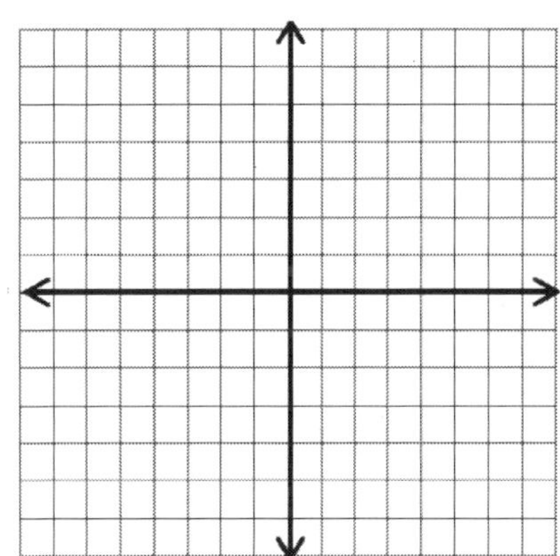

Find more at bit.ly/3hfdnJL

Writing Linear Equations

- The equation of a line in slope-intercept form: $y = mx + b$
- To write the equation of a line, first identify the slope.
- Find the y-intercept. This can be done by substituting the slope and the coordinates of a point (x, y) on the line.

Examples:

Example 1. What is the equation of the line that passes through $(3, -4)$ and has a slope of 6?

Solution: The general slope-intercept form of the equation of a line is $y = mx + b$, where m is the slope and b is the y-intercept.

By substitution of the given point and given slope:

$y = mx + b \rightarrow -4 = (6)(3) + b$. So, $b = -4 - 18 = -22$, and the required equation of the line is: $y = 6x - 22$

Example 2. Write the equation of the line through two points $A(3,1)$ and $B(-2,6)$.

Solution: First, find the slope: $Slop = \frac{y_2 - y_1}{x_2 - x_1} = \frac{6-1}{-2-3} = \frac{5}{-5} = -1 \rightarrow m = -1$

To find the value of b, use either points and plug in the values of x and y in the equation. The answer will be the same: $y = -x + b$. Let's check both points. Then:
$(3,1) \rightarrow y = mx + b \rightarrow 1 = -1(3) + b \rightarrow b = 4$

$(-2,6) \rightarrow y = mx + b \rightarrow 6 = -1(-2) + b \rightarrow b = 4$

The y-intercept of the line is 4. The equation of the line is: $y = -x + 4$

Practices:

✍ ***Write the equation of the line through the given points.***

1) through: $(1, 2), (2, 3)$
2) through: $(1, 7), (-1, 3)$
3) through: $(2, 3), (4, 4)$
4) through: $(0, 6), (1, 3)$
5) through: $(1, 4), (2, 3)$
6) through: $(1, -1), (-1, 2)$

Finding Midpoint

- The middle of a line segment is its midpoint.
- The Midpoint of two endpoints A (x_1, y_1) and B (x_2, y_2) can be found using this formula: M $\left(\frac{x_1+x_2}{2}, \frac{y_1+y_2}{2}\right)$

Examples:

Example 1. Find the midpoint of the line segment with the given endpoints. $(2, -4), (6, 8)$

Solution: Midpoint $= \left(\frac{x_1+x_2}{2}, \frac{y_1+y_2}{2}\right) \rightarrow (x_1, y_1) = (2, -4)$ and $(x_2, y_2) = (6, 8)$

Midpoint $= \left(\frac{2+6}{2}, \frac{-4+8}{2}\right) \rightarrow \left(\frac{8}{2}, \frac{4}{2}\right) \rightarrow M(4, 2)$

Find the midpoint of the line segment with the given endpoints. $(-2, 3), (6, -7)$

Solution: Midpoint $= \left(\frac{x_1+x_2}{2}, \frac{y_1+y_2}{2}\right) \rightarrow (x_1, y_1) = (-2, 3)$ and $(x_2, y_2) = (6, -7)$

Midpoint $= \left(\frac{-2+6}{2}, \frac{3+(-7)}{2}\right) \rightarrow \left(\frac{4}{2}, \frac{-4}{2}\right) \rightarrow M(2, -2)$

Practices:

Find the midpoint of the line segment with the given endpoints.

1) $(-4, -6), (0, 2)$
2) $(-2, 1), (-2, 5)$
3) $(0, -1), (2, -3)$
4) $(7, 0), (-1, 2)$
5) $(4, -3), (8, -7)$
6) $(-2, 3), (2, 5)$
7) $(1, 0), (-5, 6)$
8) $(-8, 4), (-4, 0)$
9) $(-7, 3), (9, 1)$
10) $(2, 7), (6, 9)$

bit.ly/3nPdnTq
Find more at

Finding Distance of Two Points

- Use the following formula to find the distance of two points with the coordinates A (x_1, y_1) and B (x_2, y_2):

$$d = \sqrt{(x_2 - x_1)^2 + (y_2 - y_1)^2}$$

Examples:

Example 1. Find the distance between $(4, 2)$ and $(-5, -10)$ on the coordinate plane.

Solution: Use distance of two points formula: $d = \sqrt{(x_2 - x_1)^2 + (y_2 - y_1)^2}$

$(x_1, y_1) = (4, 2)$ and $(x_2, y_2) = (-5, -10)$. Then: $d = \sqrt{(x_2 - x_1)^2 + (y_2 - y_1)^2} \rightarrow$

$$= \sqrt{(-5 - 4)^2 + (-10 - 2)^2} = \sqrt{(-9)^2 + (-12)^2} = \sqrt{81 + 144} = \sqrt{225} = 15$$

Then: $d = 15$

Example 2. Find the distance of two points $(-1, 5)$ and $(-4, 1)$.

Solution: Use distance of two points formula: $d = \sqrt{(x_2 - x_1)^2 + (y_2 - y_1)^2}$

$(x_1, y_1) = (-1, 5)$ and $(x_2, y_2) = (-4, 1)$. Then: $= \sqrt{(x_2 - x_1)^2 + (y_2 - y_1)^2} \rightarrow$

$$d = \sqrt{(-4 - (-1))^2 + (1 - 5)^2} = \sqrt{(-3)^2 + (-4)^2} = \sqrt{9 + 16} = \sqrt{25} = 5$$

Then: $d = 5$

Practices:

Find the distance between each pair of points.

1) $(3, 4), (7, 7)$

2) $(8, 18), (2, 10)$

3) $(12, -8), (0, 1)$

4) $(4, 10), (-4, -5)$

5) $(-6, 11), (10, -1)$

6) $(6, 10), (-6, 5)$

Graphing Linear Inequalities

- To graph a linear inequality, first draw a graph of the "equals" line.
- Use a dash line for less than (<) and greater than (>) signs and a solid line for less than and equal to (≤) and greater than and equal to (≥).
- Choose a testing point. (it can be any point on both sides of the line.)
- Put the value of (x, y) of that point in the inequality. If that works, that part of the line is the solution. If the values don't work, then the other part of the line is the solution.

Example:

Sketch the graph of inequality: $y < 2x + 4$

Solution: To draw the graph of $y < 2x + 4$, you first need to graph the line:

$y = 2x + 4$

Since there is a less than (<) sign, draw a dash line.

The slope is 2 and the y-intercept is 4.

Then, choose a testing point and substitute the value of x and y from that point into the inequality. The easiest point to test is the origin: $(0, 0)$

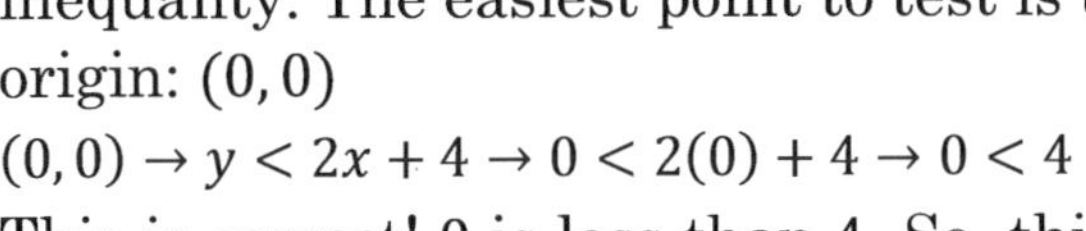

$(0,0) \rightarrow y < 2x + 4 \rightarrow 0 < 2(0) + 4 \rightarrow 0 < 4$

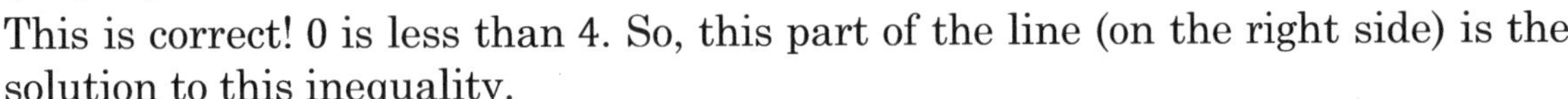

This is correct! 0 is less than 4. So, this part of the line (on the right side) is the solution to this inequality.

Practices:

Sketch the graph of each linear inequality.

1) $y < \frac{3}{4}x + 3$

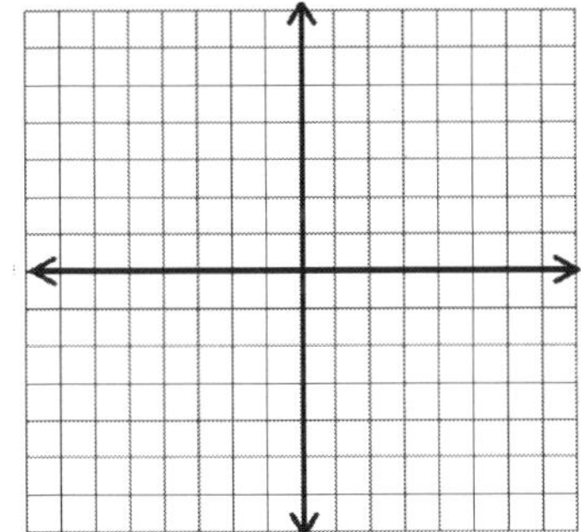

2) $y < -\frac{3}{2}x + 3$

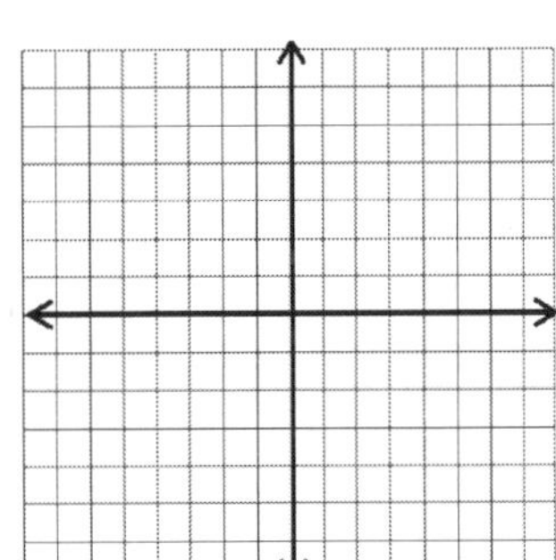

bit.ly/3rG8FsX
Find more at

Chapter 9: Answers-

Finding Slope

1) 4 ($Slop = \frac{y_2 - y_1}{x_2 - x_1}$. Let (x_1, y_1) be A $(3,2)$ and (x_2, y_2) be $B(4,6)$. Then: $Slop = \frac{y_2 - y_1}{x_2 - x_1} = \frac{6-2}{4-3} = \frac{4}{1} = 4$)

2) 3 ($Slop = \frac{y_2 - y_1}{x_2 - x_1}$. Let (x_1, y_1) be A $(2,1)$ and (x_2, y_2) be $B(3,4)$. Then: $Slop = \frac{y_2 - y_1}{x_2 - x_1} = \frac{4-1}{3-2} = \frac{3}{1} = 3$)

3) -3 (Slope $= \frac{y_2 - y_1}{x_2 - x_1}$. Let (x_1, y_1) be A $(0,3)$ and (x_2, y_2) be $B\ (1,0)$. Then: $Slop = \frac{y_2 - y_1}{x_2 - x_1} = \frac{0-3}{1-0} = \frac{-3}{1} = -3$)

4) $\frac{1}{2}$ ($Slop = \frac{y_2 - y_1}{x_2 - x_1}$. Let (x_1, y_1) be A $(4,3)$ and (x_2, y_2) be $B\ (8,5)$. Then: $Slop = \frac{y_2 - y_1}{x_2 - x_1} = \frac{5-3}{8-4} = \frac{2}{4} = \frac{1}{2}$)

5) -5 ($Slop = \frac{y_2 - y_1}{x_2 - x_1}$. Let (x_1, y_1) be A $(9,8)$ and (x_2, y_2) be $B\ (6,23)$. Then: $Slop = \frac{y_2 - y_1}{x_2 - x_1} = \frac{23-8}{6-9} = \frac{15}{-3} = -5$)

6) $\frac{1}{2}$ ($Slop = \frac{y_2 - y_1}{x_2 - x_1}$. Let (x_1, y_1) be A $(10,8)$ and (x_2, y_2) be $B\ (12,9)$. Then: $Slop = \frac{y_2 - y_1}{x_2 - x_1} = \frac{9-8}{12-10} = \frac{1}{2}$)

7) -1 ($Slop = \frac{y_2 - y_1}{x_2 - x_1}$. Let (x_1, y_1) be A $(2,7)$ and (x_2, y_2) be $B\ (1,8)$. Then: $Slop = \frac{y_2 - y_1}{x_2 - x_1} = \frac{8-7}{1-2} = \frac{1}{-1} = -1$)

8) 4 ($Slop = \frac{y_2 - y_1}{x_2 - x_1}$. Let (x_1, y_1) be A $(-5,-2)$ and (x_2, y_2) be $B\ (-3,6)$. Then: $Slop = \frac{y_2 - y_1}{x_2 - x_1} = \frac{6-(-2)}{-3-(-5)} = \frac{6+2}{-3+5} = \frac{8}{2} = 4$)

Graphing Lines Using Slope–Intercept Form

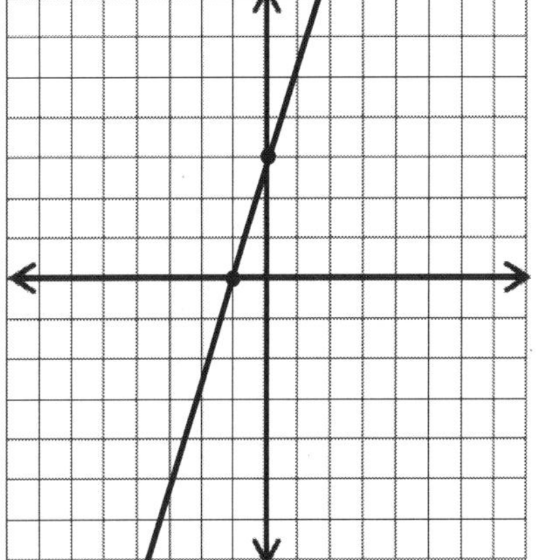

1) To graph this line, we need to find two points. When x is zero the value of y is 3. And when x is -1 the value of y is 0.
$x = 0 \rightarrow y = 3(0) + 3 = 3$
$y = 0 \rightarrow 0 = 3x + 3 \rightarrow x = -1$
Now, we have two points: $(0, 3)$ and $(-1, 0)$. Find the points on the coordinate plane and graph the line. Remember that the slope of the line is 3.

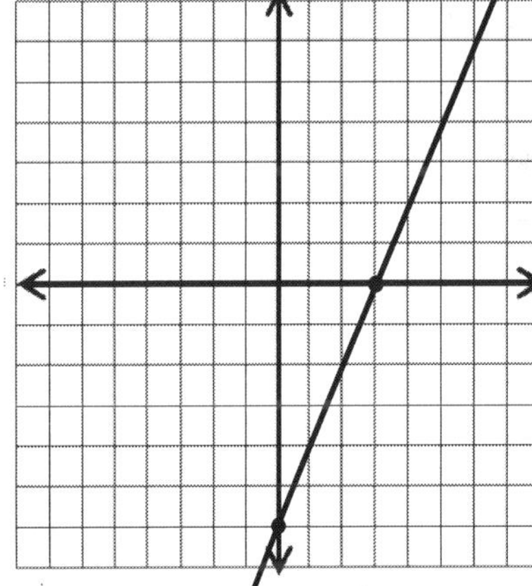

2) To graph this line, we need to find two points. When x is zero the value of y is -6. And when x is 3 the value of y is 0.
$x = 0 \rightarrow y = 2(0) - 6 = -6,$
$y = 0 \rightarrow 0 = 2x - 6 \rightarrow x = 3$
Now, we have two points: $(0, -6)$ and $(3, 0)$. Find the points on the coordinate plane and graph the line. Remember that the slope of the line is 2.

Writing Linear Equations

1) $y = x + 1$ (First, find the slope: $Slop = \frac{y_2 - y_1}{x_2 - x_1} = \frac{3-2}{2-1} = \frac{1}{1} = 1 \rightarrow m = 1$. To find the value of b, use either points and plug in the values of x and y in the equation. The answer will be the same: $y = x + b$. Then: $(1, 2) \rightarrow y = mx + b \rightarrow 2 = 1(1) + b \rightarrow b = 1$. The y-intercept of the line is 1. The equation of the line is: $y = x + 1$)

2) $y = 2x + 5$ (First, find the slope: $Slop = \frac{y_2 - y_1}{x_2 - x_1} = \frac{3-7}{-1-1} = \frac{-4}{-2} = 2 \rightarrow m = 2$. To find the value of b, use either points and plug in the values of x and y in the equation. The answer will be the same: $y = 2x + b$.

Then: $(1,7) \rightarrow y = mx + b \rightarrow 7 = 2(1) + b \rightarrow b = 5$. The y-intercept of the line is 5. The equation of the line is: $y = 2x + 5$)

3) $y = \frac{1}{2}x + 2$ (First, find the slope: $Slop = \frac{y_2 - y_1}{x_2 - x_1} = \frac{4-3}{4-2} = \frac{1}{2} \rightarrow m = \frac{1}{2}$. To find the value of b, use either points and plug in the values of x and y in the equation. The answer will be the same: $y = \frac{1}{2}x + b$. Then: $(4,4) \rightarrow y = mx + b \rightarrow 4 = \frac{1}{2}(4) + b \rightarrow b = 2$. The y-intercept of the line is 2. The equation of the line is: $y = \frac{1}{2}x + 2$)

4) $y = -3x + 6$ (First, find the slope: $Slop = \frac{y_2 - y_1}{x_2 - x_1} = \frac{3-6}{1-0} = \frac{-3}{1} = -3 \rightarrow m = -3$. To find the value of b, use either points and plug in the values of x and y in the equation. The answer will be the same: $y = -3x + b$. Then: $(0,6) \rightarrow y = mx + b \rightarrow 6 = -3(0) + b \rightarrow b = 6$. The y-intercept of the line is 6. The equation of the line is: $y = -3x + 6$)

5) $y = -x + 5$ (First, find the slope: $Slop = \frac{y_2 - y_1}{x_2 - x_1} = \frac{3-4}{2-1} = \frac{-1}{1} = -1 \rightarrow m = -1$. To find the value of b, use either points and plug in the values of x and y in the equation. The answer will be the same: $y = -x + b$. Then: $(2,3) \rightarrow y = mx + b \rightarrow 3 = -(2) + b \rightarrow b = 5$. The y-intercept of the line is 5. The equation of the line is: $y = -x + 5$)

6) $y = -\frac{3}{2}x + \frac{1}{2}$ (First, find the slope: $Slop = \frac{y_2 - y_1}{x_2 - x_1} = \frac{2-(-1)}{-1-1} = \frac{3}{-2} = -\frac{3}{2} \rightarrow m = -\frac{3}{2}$. To find the value of b, use either points and plug in the values of x and y in the equation. The answer will be the same: $y = -\frac{3}{2}x + b$. Then: $(1,-1) \rightarrow y = mx + b \rightarrow -1 = -\frac{3}{2}(1) + b \rightarrow b = \frac{1}{2}$. The y-intercept of the line is $\frac{1}{2}$. The equation of the line is: $y = -\frac{3}{2}x + \frac{1}{2}$)

Finding Midpoint

1) $M(-2,-2)$ (Midpoint $= \left(\frac{x_1+x_2}{2}, \frac{y_1+y_2}{2}\right) \rightarrow (x_1, y_1) = (-4,-6)$ and $(x_2, y_2) = (0,2)$. Midpoint $= \left(\frac{-4+0}{2}, \frac{-6+2}{2}\right) \rightarrow \left(\frac{-4}{2}, \frac{-4}{2}\right) \rightarrow M(-2,-2)$)

2) $M(-2,3)$ (Midpoint $= \left(\frac{x_1+x_2}{2}, \frac{y_1+y_2}{2}\right) \rightarrow (x_1, y_1) = (-2,1)$ and $(x_2, y_2) = (-2,5)$. Midpoint $= \left(\frac{-2+(-2)}{2}, \frac{1+5}{2}\right) \rightarrow \left(\frac{-4}{2}, \frac{6}{2}\right) \rightarrow M(-2,3)$)

3) $M(1,-2)$ (Midpoint $= \left(\frac{x_1+x_2}{2}, \frac{y_1+y_2}{2}\right) \rightarrow (x_1, y_1) = (0,-1)$ and $(x_2, y_2) = (2,-3)$. Midpoint $= \left(\frac{0+2}{2}, \frac{-1+(-3)}{2}\right) \rightarrow \left(\frac{2}{2}, \frac{-4}{2}\right) \rightarrow M(1,-2)$)

4) $M(3,1)$ (Midpoint $= \left(\frac{x_1+x_2}{2}, \frac{y_1+y_2}{2}\right) \rightarrow (x_1, y_1) = (7,0)$ and $(x_2, y_2) = (-1,2)$. Midpoint $= \left(\frac{7+(-1)}{2}, \frac{0+2}{2}\right) \rightarrow \left(\frac{6}{2}, \frac{2}{2}\right) \rightarrow M(3,1)$)

5) $M(6,-5)$ (Midpoint $= \left(\frac{x_1+x_2}{2}, \frac{y_1+y_2}{2}\right) \rightarrow (x_1, y_1) = (4,-3)$ and $(x_2, y_2) = (8,-7)$. Midpoint $= \left(\frac{4+8}{2}, \frac{-3+(-7)}{2}\right) \rightarrow \left(\frac{12}{2}, \frac{-10}{2}\right) \rightarrow M(6,-5)$)

6) $M(0,4)$ (Midpoint $= \left(\frac{x_1+x_2}{2}, \frac{y_1+y_2}{2}\right) \rightarrow (x_1, y_1) = (-2,3)$ and $(x_2, y_2) = (2,5)$. Midpoint $= \left(\frac{-2+2}{2}, \frac{3+5}{2}\right) \rightarrow \left(\frac{0}{2}, \frac{8}{2}\right) \rightarrow M(0,4)$)

7) $M(-2,3)$ (Midpoint $= \left(\frac{x_1+x_2}{2}, \frac{y_1+y_2}{2}\right) \rightarrow (x_1, y_1) = (1,0)$ and $(x_2, y_2) = (-5,6)$. Midpoint $= \left(\frac{1+(-5)}{2}, \frac{0+6}{2}\right) \rightarrow \left(\frac{-4}{2}, \frac{6}{2}\right) \rightarrow M(-2,3)$)

8) $M(-6,2)$ (Midpoint $= \left(\frac{x_1+x_2}{2}, \frac{y_1+y_2}{2}\right) \rightarrow (x_1, y_1) = (-8,4)$ and $(x_2, y_2) = (-4,0)$. Midpoint $= \left(\frac{-8+(-4)}{2}, \frac{4+0}{2}\right) \rightarrow \left(\frac{-12}{2}, \frac{4}{2}\right) \rightarrow M(-6,2)$)

9) $M(1,2)$ (Midpoint $= \left(\frac{x_1+x_2}{2}, \frac{y_1+y_2}{2}\right) \rightarrow (x_1, y_1) = (-7,3)$ and $(x_2, y_2) = (9,1)$. Midpoint $= \left(\frac{-7+9}{2}, \frac{3+1}{2}\right) \rightarrow \left(\frac{2}{2}, \frac{4}{2}\right) \rightarrow M(1,2)$)

10) $M(4,8)$ (Midpoint $= \left(\frac{x_1+x_2}{2}, \frac{y_1+y_2}{2}\right) \rightarrow (x_1, y_1) = (2,7)$ and $(x_2, y_2) = (6,9)$. Midpoint $= \left(\frac{2+6}{2}, \frac{7+9}{2}\right) \rightarrow \left(\frac{8}{2}, \frac{16}{2}\right) \rightarrow M(4,8)$)

Finding Distance of Two Points

1) 5 (Use distance of two points formula: $d = \sqrt{(x_2 - x_1)^2 + (y_2 - y_1)^2}$

$(x_1, y_1) = (3,4)$ and $(x_2, y_2) = (7,7)$. Then:

$d = \sqrt{(x_2 - x_1)^2 + (y_2 - y_1)^2} = \sqrt{(7-3)^2 + (7-4)^2} = \sqrt{(4)^2 + (3)^2} = \sqrt{16+9} = \sqrt{25} = 5$)

2) 10 (Use distance of two points formula: $d = \sqrt{(x_2 - x_1)^2 + (y_2 - y_1)^2}$

$(x_1, y_1) = (8,18)$ and $(x_2, y_2) = (2,10)$. Then:

$d = \sqrt{(x_2 - x_1)^2 + (y_2 - y_1)^2} = \sqrt{(2-8)^2 + (10-18)^2} = \sqrt{(-6)^2 + (-8)^2} = \sqrt{36+64} = \sqrt{100} = 10$)

3) 15 (Use distance of two points formula: $d = \sqrt{(x_2 - x_1)^2 + (y_2 - y_1)^2}$

$(x_1, y_1) = (12,-8)$ and $(x_2, y_2) = (0,1)$. Then:

$d = \sqrt{(x_2 - x_1)^2 + (y_2 - y_1)^2} = \sqrt{(0-12)^2 + \left(1-(-8)\right)^2} = \sqrt{(12)^2 + (9)^2} = \sqrt{144+81} = \sqrt{225} = 15$)

4) 17 (Use distance of two points formula: $d = \sqrt{(x_2 - x_1)^2 + (y_2 - y_1)^2}$

$(x_1, y_1) = (4,10)$ and $(x_2, y_2) = (-4,-5)$. Then:

$d = \sqrt{(x_2 - x_1)^2 + (y_2 - y_1)^2} = \sqrt{(-4-4)^2 + (-5-10)^2} = \sqrt{(-8)^2 + (-15)^2} = \sqrt{64+225} = \sqrt{289} = 17$)

5) 20 (Use distance of two points formula: $d = \sqrt{(x_2 - x_1)^2 + (y_2 - y_1)^2}$

$(x_1, y_1) = (-6,11)$ and $(x_2, y_2) = (10,-1)$. Then:

$d = \sqrt{(x_2 - x_1)^2 + (y_2 - y_1)^2} = \sqrt{(10 - (-6))^2 + (-1 - 11)^2} = \sqrt{(16)^2 + (-12)^2} = \sqrt{256 + 144} = \sqrt{400} = 20$)

6) 13 (Use distance of two points formula: $d = \sqrt{(x_2 - x_1)^2 + (y_2 - y_1)^2}$

$(x_1, y_1) = (6, 10)$ and $(x_2, y_2) = (-6, 5)$. Then:

$d = \sqrt{(x_2 - x_1)^2 + (y_2 - y_1)^2} = \sqrt{(-6 - 6)^2 + (5 - 10)^2} = \sqrt{(-12)^2 + (-5)^2} = \sqrt{144 + 25} = \sqrt{169} = 13$)

Graphing Linear Inequalities

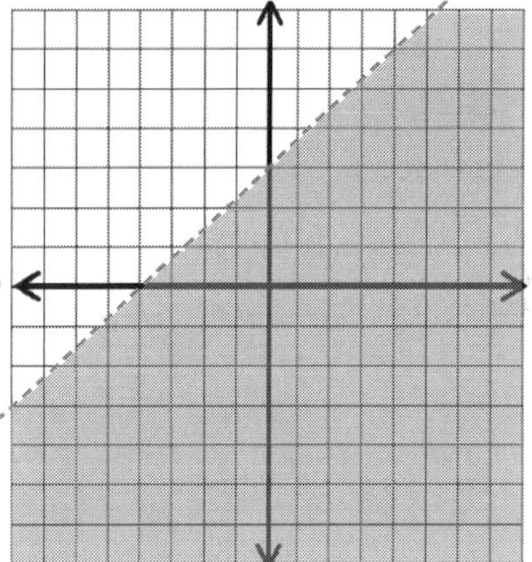

1) To draw the graph of $y < \frac{3}{4}x + 3$, you first need to graph the line: $y = \frac{3}{4}x + 3$
Since there is a less than (<) sign, draw a dash line.
The slope is $\frac{3}{4}$ and y-intercept is 3.
Then, choose a testing point and substitute the value of x and y from that point into the inequality. The easiest point to test is the origin: $(0, 0)$
$(0,0) \rightarrow y < \frac{3}{4}x + 3 \rightarrow 0 < \frac{3}{4}(0) + 3 \rightarrow 0 < 3$
This is correct! 0 is less than 3. So, this part of the line (on the right side) is the solution of this inequality.

2) To draw the graph of $y < -\frac{3}{2}x + 3$, you first need to graph the line: $y = -\frac{3}{2}x + 3$

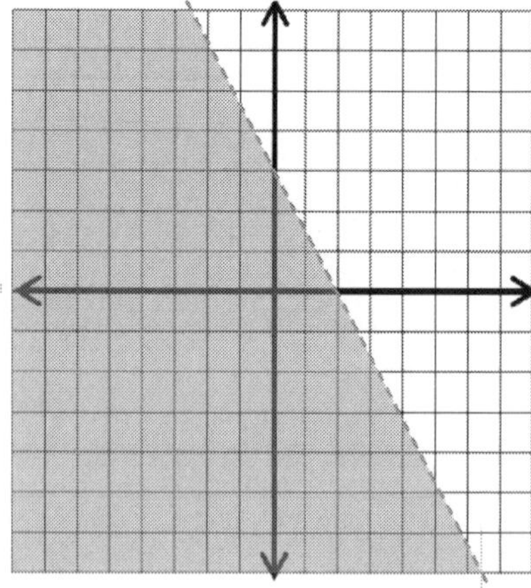

Since there is a less than (<) sign, draw a dash line.

The slope is $-\frac{3}{2}$ and y-intercept is 3.

Then, choose a testing point and substitute the value of x and y from that point into the inequality. The easiest point to test is the origin: $(0, 0)$

$$(0,0) \rightarrow y < -\frac{3}{2}x + 3 \rightarrow 0 < -\frac{3}{2}(0) + 3 \rightarrow 0 < 3$$

This is correct! 0 is less than 3. So, this part of the line (on the left side) is the solution of this inequality.

CHAPTER

10 Polynomials

Math topics that you'll learn in this chapter:

- ☑ Simplifying Polynomials
- ☑ Adding and Subtracting Polynomials
- ☑ Multiplying Monomials
- ☑ Multiplying and Dividing Monomials
- ☑ Multiplying a Polynomial and a Monomial
- ☑ Multiplying Binomials
- ☑ Factoring Trinomials

Simplifying Polynomials

- To simplify Polynomials, find "like" terms. (they have the same variables with the same power).
- Use "FOIL". (First–Out–In–Last) for binomials:

$$(x + a)(x + b) = x^2 + (b + a)x + ab$$

- Add or Subtract "like" terms using order of operation.

Examples:

Example 1. Simplify this expression. $x(4x + 7) - 2x =$

Solution: Use Distributive Property: $x(4x + 7) = 4x^2 + 7x$

Now, combine like terms: $x(4x + 7) - 2x = 4x^2 + 7x - 2x = 4x^2 + 5x$

Example 2. Simplify this expression. $(x + 3)(x + 5) =$

Solution: First, apply the FOIL method: $(a + b)(c + d) = ac + ad + bc + bd$

$(x + 3)(x + 5) = x^2 + 5x + 3x + 15$

Now combine like terms: $x^2 + 5x + 3x + 15 = x^2 + 8x + 15$

Practices:

Simplify each expression

1) $6(3x - 2) =$

2) $8x(2x + 5) =$

3) $2x(10x - 4) =$

4) $7x(4x + 3) =$

5) $11x(5x - 6) =$

6) $3x(3x + 7) =$

7) $(4x - 5)(x - 2) =$

8) $(x - 3)(10x + 7) =$

Adding and Subtracting Polynomials

- Adding polynomials is just a matter of combining like terms, with some order of operations considerations thrown in.
- Be careful with the minus signs, and don't confuse addition and multiplication!
- For subtracting polynomials, sometimes you need to use the Distributive Property: $a(b + c) = ab + ac$, $a(b - c) = ab - ac$

Examples:

Example 1. Simplify the expressions. $(x^2 - 2x^3) - (x^3 - 3x^2) =$

Solution: First, use Distributive Property: $-(x^3 - 3x^2) = -x^3 + 3x^2 \rightarrow$
$(x^2 - 2x^3) - (x^3 - 3x^2) = x^2 - 2x^3 - x^3 + 3x^2$
Now combine like terms: $-2x^3 - x^3 = -3x^3$ and $x^2 + 3x^2 = 4x^2$
Then: $(x^2 - 2x^3) - (x^3 - 3x^2) = x^2 - 2x^3 - x^3 + 3x^2 = -3x^3 + 4x^2$

Example 2. Add expressions. $(3x^3 - 5) + (4x^3 - 2x^2) =$

Solution: Remove parentheses:

$$(3x^3 - 5) + (4x^3 - 2x^2) = 3x^3 - 5 + 4x^3 - 2x^2$$

Now combine like terms: $3x^3 - 5 + 4x^3 - 2x^2 = 7x^3 - 2x^2 - 5$

Practices:

Add or subtract expressions.

1) $(-4x^2 - 3) + (6x^2 + 5) =$
2) $(x^2 + 2) - (6 - 2x^2) =$
3) $(8x^3 + x^2) - (3x^3 + 8) =$
4) $(11x^3 - 5x^2) + (3x^2 - x) =$
5) $(5x^3 + x) - (7x^3 - 6) =$
6) $(2x^3 - 12) + (9x^3 + 3) =$
7) $(6x^3 + 4) - (11 - 2x^3) =$
8) $(10x^2 + 8x^3) - (2x^3 + 7) =$

bit.ly/2KUqHqQ
Find more at

Multiplying Monomials

- A monomial is a polynomial with just one term: Examples: $2x$ or $7y^2$.
- When you multiply monomials, first multiply the coefficients (a number placed before and multiplying the variable) and then multiply the variables using the multiplication property of exponents.

$$x^a \times x^b = x^{a+b}$$

Examples:

Example 1. Multiply expressions. $2xy^3 \times 6x^4y^2$

Solution: Find the same variables and use multiplication property of exponents:
$x^a \times x^b = x^{a+b}$
$x \times x^4 = x^{1+4} = x^5$ and $y^3 \times y^2 = y^{3+2} = y^5$
Then, multiply coefficients and variables: $2xy^3 \times 6x^4y^2 = 12x^5y^5$

Example 2. Multiply expressions. $7a^3b^8 \times 3a^6b^4 =$

Solution: Use the multiplication property of exponents: $x^a \times x^b = x^{a+b}$
$a^3 \times a^6 = a^{3+6} = a^9$ and $b^8 \times b^4 = b^{8+4} = b^{12}$
Then: $7a^3b^8 \times 3a^6b^4 = 21a^9b^{12}$

Practices:

Simplify each expression

1) $4x^5 \times (-7x^4) =$

2) $6ab^7c^3 \times 3b^6 =$

3) $(-5u^8t^2) \times (-2u^2t^3) =$

4) $9x^4y^3 \times 6x^5y^{11} =$

5) $(-3p^9q^5) \times (5p^4q^4) =$

6) $11a^9b^2 \times 2a^3b^7 =$

7) $-4u^6t^2 \times 9u^8t^{12} =$

8) $(-p^{16}q^8) \times (-2pq^2) =$

Find more at bit.ly/2KLVoP8

Multiplying and Dividing Monomials

- When you divide or multiply two monomials, you need to divide or multiply their coefficients and then divide or multiply their variables.
- In the case of exponents with the same base, for Division, subtract their powers, for Multiplication, add their powers.
- Exponent's Multiplication and Division rules:

$$x^a \times x^b = x^{a+b}, \qquad \frac{x^a}{x^b} = x^{a-b}$$

Examples:

Example 1. Multiply expressions. $(3x^5)(9x^4) =$

Solution: Use multiplication property of exponents:

$x^a \times x^b = x^{a+b} \rightarrow x^5 \times x^4 = x^9$

Then: $(3x^5)(9x^4) = 27x^9$

Example 2. Divide expressions. $\frac{12x^4y^6}{6xy^2} =$

Solution: Use division property of exponents:

$\frac{x^a}{x^b} = x^{a-b} \rightarrow \frac{x^4}{x} = x^{4-1} = x^3$ and $\frac{y^6}{y^2} = y^{6-2} = y^4$

Then: $\frac{12x^4y^6}{6xy^2} = 2x^3y^4$

Practices:

Simplify each expression

1) $(5x^6y^2)(4x^5y^3) =$

2) $(-6x^7y^{10})(-8x^3y^4) =$

3) $(2xy^9)(-9x^8y) =$

4) $\frac{81x^8y^9}{9x^4y^5} =$

5) $\frac{60x^7y^5}{5x^5y^4} =$

6) $\frac{14x^7y^6}{7x^2y^3} =$

Multiplying a Polynomial and a Monomial

- When multiplying monomials, use the product rule for exponents.

$$x^a \times x^b = x^{a+b}$$

- When multiplying a monomial by a polynomial, use the distributive property.

$$a \times (b + c) = a \times b + a \times c = ab + ac$$
$$a \times (b - c) = a \times b - a \times c = ab - ac$$

Examples:

Example 1. Multiply expressions. $6x(2x + 5)$

Solution: Use Distributive Property:

$6x(2x + 5) = 6x \times 2x + 6x \times 5 = 12x^2 + 30x$

Example 2. Multiply expressions. $x(3x^2 + 4y^2)$

Solution: Use Distributive Property:

$x(3x^2 + 4y^2) = x \times 3x^2 + x \times 4y^2 = 3x^3 + 4xy^2$

Practices:

Find each product.

1) $6x(10x + 5y) =$
2) $10x(6x - 3y) =$
3) $7x(9x + 2y) =$
4) $11x(x + 10) =$
5) $4x(-9x^2y + 3y) =$
6) $-3x(4x^2 + 5xy - 9) =$
7) $8x(7x - 2y + 8) =$
8) $6x(4x^2 + 6y^2) =$

Multiplying Binomials

- A binomial is a polynomial that is the sum or the difference of two terms, each of which is a monomial.
- To multiply two binomials, use the "FOIL" method. (First–Out–In–Last)

$$(x+a)(x+b) = x \times x + x \times b + a \times x + a \times b = x^2 + bx + ax + ab$$

Examples:

Example 1. Multiply Binomials. $(x+3)(x-2) =$

Solution: Use "FOIL". (First–Out–In–Last):

$(x+3)(x-2) = x^2 - 2x + 3x - 6$

Then combine like terms: $x^2 - 2x + 3x - 6 = x^2 + x - 6$

Example 2. Multiply. $(x+6)(x+4) =$

Solution: Use "FOIL". (First–Out–In–Last):

$(x+6)(x+4) = x^2 + 4x + 6x + 24$

Then simplify: $x^2 + 4x + 6x + 24 = x^2 + 10x + 24$

Practices:

Find each product.

1) $(x+5)(x+5) =$
2) $(x+5)(x-10) =$
3) $(x+8)(x+9) =$
4) $(x-7)(x-8) =$
5) $(x+1)(x+4) =$
6) $(x+6)(x-9) =$
7) $(x-3)(x+7) =$
8) $(x-6)(x-4) =$

Factoring Trinomials

To factor trinomials, you can use following methods:

- "FOIL": $(x+a)(x+b) = x^2 + (b+a)x + ab$
- "Difference of Squares":

$$a^2 - b^2 = (a+b)(a-b)$$
$$a^2 + 2ab + b^2 = (a+b)(a+b)$$
$$a^2 - 2ab + b^2 = (a-b)(a-b)$$

- "Reverse FOIL": $x^2 + (b+a)x + ab = (x+a)(x+b)$

Examples:

Example 1. Factor this trinomial. $x^2 - 2x - 8$

Solution: Break the expression into groups. You need to find two numbers that their product is -8 and their sum is -2. (remember "Reverse FOIL": $x^2 + (b+a)x + ab = (x+a)(x+b)$). Those two numbers are 2 and -4. Then:

$$x^2 - 2x - 8 = (x^2 + 2x) + (-4x - 8)$$

Now factor out x from $x^2 + 2x$: $x(x+2)$, and factor out -4 from $-4x - 8$: $-4(x+2)$; Then: $(x^2 + 2x) + (-4x - 8) = x(x+2) - 4(x+2)$

Now factor out like term: $(x+2)$. Then: $(x+2)(x-4)$

Example 2. Factor this trinomial. $x^2 - 2x - 24$

Solution: Break the expression into groups: $(x^2 + 4x) + (-6x - 24)$

Now factor out x from $x^2 + 4x$: $x(x+4)$, and factor out -6 from $-6x - 24$: $-6(x+4)$; Then: $x(x+4) - 6(x+4)$, now factor out like term: $(x+4) \rightarrow x(x+4) - 6(x+4) = (x+4)(x-6)$

Practices:

Factor each trinomial.

1) $x^2 + 11x + 30 =$
2) $x^2 - 6x - 7 =$
3) $x^2 + 8x + 7 =$
4) $x^2 - 12x + 20 =$
5) $x^2 - x - 20 =$
6) $x^2 - 3x - 54 =$
7) $x^2 + 21x + 110 =$
8) $x^2 - 9x - 36 =$

Chapter 10: Answers

Simplifying Polynomials

1) $18x - 12$ (Use Distributive Property: $6(3x - 2) = (6 \times 3x) - (6 \times 2)$. Then, $6(3x - 2) = 18x - 12$)

2) $16x^2 + 40x$ (Use Distributive Property: $8x(2x + 5) = (8x \times 2x) + (8x \times 5)$. Then, $8x(2x + 5) = 16x^2 + 40x$)

3) $20x^2 - 8x$ (Use Distributive Property: $2x(10x - 4) = (2x \times 10x) - (2x \times 4)$. Then, $2x(10x - 4) = 20x^2 - 8x$)

4) $28x^2 + 21x$ (Use Distributive Property: $7x(4x + 3) = (7x \times 4x) + (7x \times 3)$. Then, $7x(4x + 3) = 28x^2 + 21x$)

5) $55x^2 - 66x$(Use Distributive Property: $11x(5x - 6) = (11x \times 5x) - (11x \times 6)$. Then, $11x(5x - 6) = 55x^2 - 66x$)

6) $9x^2 + 21x$ (Use Distributive Property: $3x(3x + 7) = (3x \times 3x) + (3x \times 7)$. Then, $3x(3x + 7) = 9x^2 + 21x$)

7) $4x^2 - 13x + 10$ (First, apply the FOIL method: $(a + b)(c + d) = ac + ad + bc + bd \rightarrow (4x - 5)(x - 2) = 4x^2 - 8x - 5x + 10$. Now combine like terms: $4x^2 - 8x - 5x + 10 = 4x^2 - 13x + 10$)

8) $10x^2 - 23x - 21$ (First, apply the FOIL method: $(a + b)(c + d) = ac + ad + bc + bd \rightarrow (x - 3)(10x + 7) = 10x^2 + 7x - 30x - 21$. Now combine like terms: $10x^2 + 7x - 30x - 21 = 10x^2 - 23x - 21$)

Adding and Subtracting Polynomials

1) $2x^2 + 2$ (Remove parentheses:$(-4x^2 - 3) + (6x^2 + 5) = -4x^2 - 3 + 6x^2 + 5$. Now combine like terms: $-4x^2 - 3 + 6x^2 + 5 = 2x^2 + 2$)

2) $3x^2 - 4$ (First, use Distributive Property: $-(6 - 2x^2) = -6 + 2x^2 \rightarrow (x^2 + 2) - (6 - 2x^2) = x^2 + 2 - 6 + 2x^2$. Now combine like terms:

$x^2 + 2x^2 = 3x^2$ and $2 - 6 = -4$. Then: $(x^2 + 2) - (6 - 2x^2) = 3x^2 - 4$)

3) $5x^3 + x^2 - 8$ (First, use Distributive Property: $-(3x^3 + 8) = -3x^3 - 8 \rightarrow$ $(8x^3 + x^2) - (3x^3 + 8) = 8x^3 + x^2 - 3x^3 - 8$. Now combine like terms: $8x^3 - 3x^3 = 5x^3$. Then: $(8x^3 + x^2) - (3x^3 + 8) = 5x^3 + x^2 - 8$)

4) $11x^3 - 2x^2 - x$ (Remove parentheses:$(11x^3 - 5x^2) + (3x^2 - x) =$ $11x^3 - 5x^2 + 3x^2 - x$. Now combine like terms: $11x^3 - 5x^2 + 3x^2 - x =$ $11x^3 - 2x^2 - x$)

5) $-2x^3 + x + 6$ (First, use Distributive Property: $-(7x^3 - 6) = -7x^3 + 6 \rightarrow$ $(5x^3 + x) - (7x^3 - 6) = 5x^3 + x - 7x^3 + 6$. Now combine like terms: $5x^3 + x - 7x^3 + 6 = -2x^3 + x + 6$. Then: $(5x^3 + x) - (7x^3 - 6) =$ $-2x^3 + x + 6$)

6) $11x^3 - 9$ (Remove parentheses:$(2x^3 - 12) + (9x^3 + 3) =$ $2x^3 - 12 + 9x^3 + 3$. Now combine like terms: $2x^3 - 12 + 9x^3 + 3 =$ $11x^3 - 9$)

7) $8x^3 - 7$ (First, use Distributive Property: $-(11 - 2x^3) = -11 + 2x^3 \rightarrow$ $(6x^3 + 4) - (11 - 2x^3) = 6x^3 + 4 - 11 + 2x^3$. Now combine like terms: $6x^3 + 4 - 11 + 2x^3 = 8x^3 - 7$. Then: $(6x^3 + 4) - (11 - 2x^3) = 8x^3 - 7$)

8) $6x^3 + 10x^2 - 7$ (First, use Distributive Property: $-(2x^3 + 7) = -2x^3 - 7 \rightarrow$ $(10x^2 + 8x^3) - (2x^3 + 7) = 10x^2 + 8x^3 - 2x^3 - 7$. Now combine like terms and write in standard form: $10x^2 + 8x^3 - 2x^3 - 7 = 6x^3 + 10x^2 - 7$)

Multiplying Monomials

1) $-28x^9$ (Use the multiplication property of exponents: $x^a \times x^b = x^{a+b} \rightarrow$ $x^5 \times x^4 = x^{5+4} = x^9$. Then: $4x^5 \times (-7x^4) = -28x^9$)

2) $18ab^{13}c^3$ (Find the same variables and use multiplication property of exponents: $x^a \times x^b = x^{a+b} \rightarrow b^7 \times b^6 = b^{7+6} = b^{13}$. Then, multiply coefficients and variables: $6ab^7c^3 \times 3b^6 = 18ab^{13}c^3$)

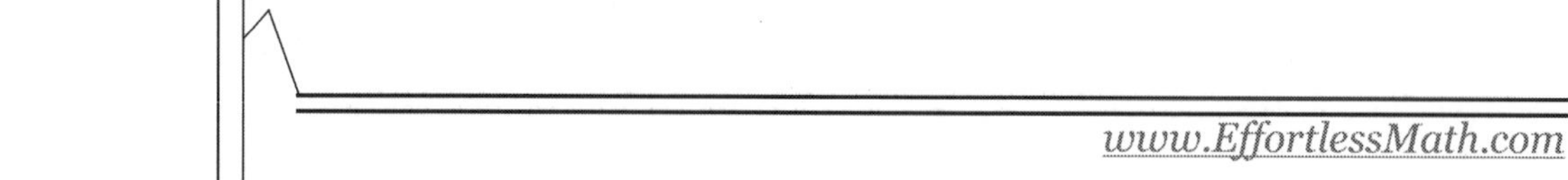

3) $10u^{10}t^5$(Use the multiplication property of exponents: $x^a \times x^b = x^{a+b} \rightarrow u^8 \times u^2 = u^{8+2} = u^{10}$ and $t^2 \times t^3 = t^{2+3} = t^5$. Then:

$(-5u^8t^2) \times (-2u^2t^3) = 10u^{10}t^5$)

4) $54x^9y^{14}$ (Use the multiplication property of exponents: $x^a \times x^b = x^{a+b} \rightarrow x^4 \times x^5 = x^{4+5} = x^9$ and $y^3 \times y^{11} = y^{3+11} = y^{14}$. Then: $9x^4y^3 \times 6x^5y^{11} = 54x^9y^{14}$)

5) $-15p^{13}q^9$(Use the multiplication property of exponents: $x^a \times x^b = x^{a+b} \rightarrow p^9 \times p^4 = p^{9+4} = p^{13}$ and $q^5 \times q^4 = q^{5+4} = q^9$. Then: $(-3p^9q^5) \times (5p^4q^4) = -15p^{13}q^9$)

6) $22a^{12}b^9$(Use the multiplication property of exponents: $x^a \times x^b = x^{a+b} \rightarrow a^9 \times a^3 = a^{9+3} = a^{12}$ and $b^2 \times b^7 = b^{2+7} = b^9$. Then: $11a^9b^2 \times 2a^3b^7 = 22a^{12}b^9$)

7) $-36u^{14}t^{14}$ (Use the multiplication property of exponents: $x^a \times x^b = x^{a+b} \rightarrow u^6 \times u^8 = u^{6+8} = u^{14}$ and $t^2 \times t^{12} = t^{2+12} = t^{14}$. Then: $-4u^6t^2 \times 9u^8t^{12} = -36u^{14}t^{14}$)

8) $2p^{17}q^{10}$(Use the multiplication property of exponents: $x^a \times x^b = x^{a+b} \rightarrow p^{16} \times p = p^{16+1} = p^{17}$ and $q^8 \times q^2 = q^{8+2} = q^{10}$. Then: $(-p^{16}q^8) \times (-2pq^2) = 2p^{17}q^{10}$)

Multiplying and Dividing Monomials

1) $20x^{11}y^5$ (Use multiplication property of exponents: $x^a \times x^b = x^{a+b} \rightarrow x^6 \times x^5 = x^{11}$ and$y^2 \times y^3 = y^5$. Then: $(5x^6y^2)(4x^5y^3) = 20x^{11}y^5$)

2) $48x^{10}y^{14}$ (Use multiplication property of exponents: $x^a \times x^b = x^{a+b} \rightarrow x^7 \times x^3 = x^{10}$ and $y^{10} \times y^4 = y^{14}$. Then: $(-6x^7y^{10})(-8x^3y^4) = 48x^{10}y^{14}$)

3) $-18x^9y^{10}$ (Use multiplication property of exponents: $x^a \times x^b = x^{a+b} \rightarrow x \times x^8 = x^9$ and$y^9 \times y = y^{10}$. Then: $(2xy^9)(-9x^8y) = -18x^9y^{10}$)

4) $9x^4y^4$ (Use division property of exponents: $\frac{x^a}{x^b} = x^{a-b} \rightarrow$

$\frac{x^8}{x^4} = x^{8-4} = x^4$ and $\frac{y^9}{y^5} = y^{9-5} = y^4$. Then: $\frac{81x^8y^9}{9x^4y^5} = 9x^4y^4$)

5) $12x^2y$ (Use division property of exponents: $\frac{x^a}{x^b} = x^{a-b} \rightarrow \frac{x^7}{x^5} = x^{7-5} = x^2$ and $\frac{y^5}{y^4} = y^{5-4} = y$. Then: $\frac{60x^7y^5}{5x^5y^4} = 12x^2y$)

6) $2x^5y^3$ (Use division property of exponents: $\frac{x^a}{x^b} = x^{a-b} \rightarrow \frac{x^7}{x^2} = x^{7-2} = x^5$ and $\frac{y^6}{y^3} = y^{6-3} = y^3$. Then: $\frac{14x^7y^6}{7x^2y^3} = 2x^5y^3$)

Multiplying a Polynomial and a Monomial

1) $60x^2 + 30xy$ (Use Distributive Property: $6x(10x + 5y) = 6x \times 10x + 6x \times 5y = 60x^2 + 30xy$)

2) $60x^2 - 30xy$ (Use Distributive Property: $10x(6x - 3y) = 10x \times 6x + 10x \times (-3y) = 60x^2 - 30xy$)

3) $63x^2 + 14xy$ (Use Distributive Property: $7x(9x + 2y) = 7x \times 9x + 7x \times 2y = 63x^2 + 14xy$)

4) $11x^2 + 110x$ (Use Distributive Property: $11x(x + 10) = 11x \times x + 11x \times 10 = 11x^2 + 110x$)

5) $-36x^3y + 12xy$ (Use Distributive Property: $4x(-9x^2y + 3y) = 4x \times (-9x^2y) + 4x \times 3y = -36x^3y + 12xy$)

6) $-12x^3 - 15x^2y + 27x$ (Use Distributive Property: $-3x(4x^2 + 5xy - 9) = -3x \times 4x^2 + (-3x) \times 5xy + (-3x)(-9) = -12x^3 - 15x^2y + 27x$)

7) $56x^2 - 16xy + 64x$ (Use Distributive Property: $8x(7x - 2y + 8) = 8x \times 7x + 8x \times (-2y) + 8x \times 8 = 56x^2 - 16xy + 64x$)

8) $24x^3 + 36xy^2$ (Use Distributive Property: $6x(4x^2 + 6y^2) = 6x \times 4x^2 + 6x \times 6y^2 = 24x^3 + 36xy^2$)

Multiplying Binomials

1) $x^2 + 10x + 25$ (Use "FOIL". (First–Out–In–Last): $(x + 5)(x + 5) =$ $x^2 + 5x + 5x + 25$. Then combine like terms: $x^2 + 5x + 5x + 25 =$ $x^2 + 10x + 25$)

2) $x^2 - 5x - 50$ (Use "FOIL". (First–Out–In–Last): $(x + 5)(x - 10) =$ $x^2 - 10x + 5x - 50$. Then combine like terms: $x^2 - 10x + 5x - 50 =$ $x^2 - 5x - 50$)

3) $x^2 + 17x + 72$ (Use "FOIL". (First–Out–In–Last): $(x + 8)(x + 9) =$ $x^2 + 9x + 8x + 72$. Then combine like terms: $x^2 + 9x + 8x + 72 =$ $x^2 + 17x + 72$)

4) $x^2 - 15x + 56$ (Use "FOIL". (First–Out–In–Last): $(x - 7)(x - 8) =$ $x^2 - 8x - 7x + 56$. Then combine like terms: $x^2 - 8x - 7x + 56 =$ $x^2 - 15x + 56$)

5) $x^2 + 5x + 4$ (Use "FOIL". (First–Out–In–Last): $(x + 1)(x + 4) =$ $x^2 + 4x + x + 4$. Then combine like terms: $x^2 + 4x + x + 4 = x^2 + 5x + 4$)

6) $x^2 - 3x - 54$ (Use "FOIL". (First–Out–In–Last): $(x + 6)(x - 9) =$ $x^2 + 6x - 9x - 54$. Then combine like terms: $x^2 + 6x - 9x - 54 =$ $x^2 - 3x - 54$)

7) $x^2 + 4x - 21$ (Use "FOIL". (First–Out–In–Last): $(x - 3)(x + 7) =$ $x^2 + 7x - 3x - 21$. Then combine like terms: $x^2 + 7x - 3x - 21 =$ $x^2 + 4x - 21$)

8) $x^2 - 10x + 24$ (Use "FOIL". (First–Out–In–Last): $(x - 4)(x - 6) =$ $x^2 - 4x - 6x + 24$. Then combine like terms: $x^2 - 4x - 6x + 24 =$ $x^2 - 10x + 24$)

Factoring Trinomials

1) $(x+5)(x+6)$ (Break the expression into groups: $(x^2+6x)+(5x+30)$. Now factor out x from $x^2+6x : x(x+6)$, and factor out 5 from $5x+30$: $5(x+6)$; Then: $(x^2+6x)+(5x+30)=x(x+6)+5(x+6)$, now factor out like term: $(x+6) \rightarrow x(x+6)+5(x+6)=(x+5)(x+6)$)

2) $(x-7)(x+1)$ (Break the expression into groups: $(x^2-7x)+(x-7)$. Now factor out x from $x^2-7x : x(x-7)$, Then: $(x^2-7x)+(x-7)= x(x-7)+(x-7)$, now factor out like term: $(x-7) \rightarrow x(x-7)+(x-7)=(x-7)(x+1)$)

3) $(x+7)(x+1)$ (Break the expression into groups: $(x^2+7x)+(x+7)$. Now factor out x from $x^2+7x : x(x+7)$, Then: $(x^2+7x)+(x+7)= x(x+7)+(x+7)$, now factor out like term: $(x+7) \rightarrow x(x+7)+(x+7)=(x+7)(x+1)$)

4) $(x-2)(x-10)$ (Break the expression into groups: $(x^2-2x)+(-10x+20)$. Now factor out x from $x^2-2x : x(x-2)$, and factor out -10 from $(-10x+20)$: $-10(x-2)$;Then: $(x^2-2x)+(-10x+20)= x(x-2)-10(x-2)$, now factor out like term: $(x-2) \rightarrow x(x-2)-10(x-2)=(x-2)(x-10)$)

5) $(x-5)(x+4)$ (Break the expression into groups: $(x^2-5x)+(4x-20)$. Now factor out x from $x^2-5x : x(x-5)$, and factor out 4 from $4x-20$: $4(x-5)$;Then: $(x^2-5x)+(4x-20)=x(x-5)+4(x-5)$, now factor out like term: $(x-5) \rightarrow x(x-5)+4(x-5)=(x-5)(x+4)$)

6) $(x-9)(x+6)$ (Break the expression into groups: $(x^2-9x)+(6x-54)$. Now factor out x from $x^2-9x : x(x-9)$, and factor out 6 from $6x-54$: $6(x-9)$;Then: $(x^2-9x)+(6x-54)=x(x-9)+6(x-9)$, now factor out like term: $(x-9) \rightarrow x(x-9)+6(x-9)=(x-9)(x+6)$)

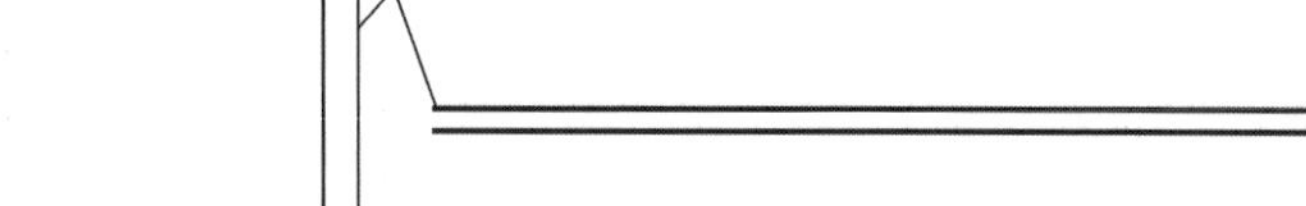

7) $(x+11)(x+10)$ (Break the expression into groups: $(x^2+11x)+(10x+110)$. Now factor out x from $x^2+11x : x(x+11)$, and factor out 10 from $10x+110$: $10(x+11)$;Then: $(x^2+11x)+(10x+110)=x(x+11)+10(x+11)$, now factor out like term: $(x+11) \rightarrow x(x+11)+10(x+11)=(x+11)(x+10)$)

8) $(x+3)(x-12)$ (Break the expression into groups: $(x^2+3x)+(-12x-36)$. Now factor out x from $x^2+3x : x(x+3)$, and factor out -12 from $(-12x-36)$: $-12(x+3)$;Then: $(x^2+3x)+(-12x-36)=x(x+3)-12(x+3)$, now factor out like term: $(x+3) \rightarrow x(x+3)-12(x+3)=(x+3)(x-12)$)

CHAPTER

11 Geometry and Solid Figures

Math topics that you'll learn in this chapter:

- ☑ The Pythagorean Theorem
- ☑ Complementary and Supplementary angles
- ☑ Parallel lines and Transversals
- ☑ Triangles
- ☑ Special Right Triangles
- ☑ Polygons
- ☑ Circles
- ☑ Trapezoids
- ☑ Cubes
- ☑ Rectangle Prisms
- ☑ Cylinder

The Pythagorean Theorem

- You can use the Pythagorean Theorem to find a missing side in a right triangle.
- In any right triangle: $a^2 + b^2 = c^2$

Examples:

Example 1. Right triangle ABC (not shown) has two legs of lengths 3 cm (AB) and 4 cm (AC). What is the length of the hypotenuse of the triangle (side BC)?

Solution: Use Pythagorean Theorem: $a^2 + b^2 = c^2$, $a = 3$, and $b = 4$

Then: $a^2 + b^2 = c^2 \rightarrow 3^2 + 4^2 = c^2 \rightarrow 9 + 16 = c^2 \rightarrow 25 = c^2 \rightarrow c = \sqrt{25} = 5$

The length of the hypotenuse is 5 cm.

Example 2. Find the hypotenuse of this triangle.

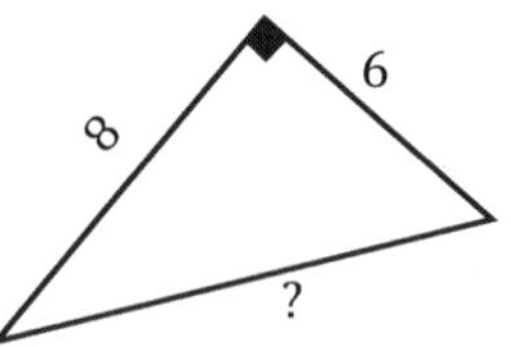

Solution: Use Pythagorean Theorem: $a^2 + b^2 = c^2$

Then: $a^2 + b^2 = c^2 \rightarrow 8^2 + 6^2 = c^2 \rightarrow 64 + 36 = c^2$

$c^2 = 100 \rightarrow c = \sqrt{100} = 10$

Practices:

Find the missing side.

1)

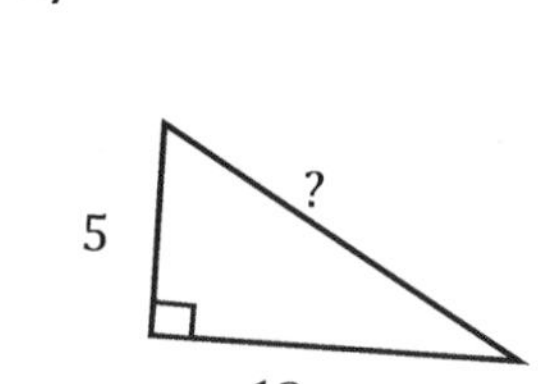

2)

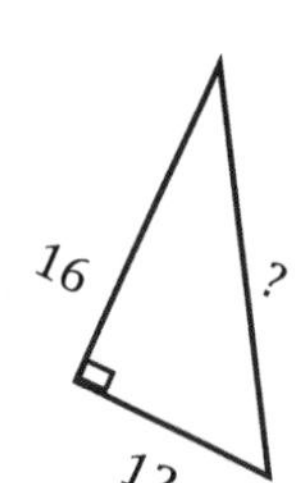

3)

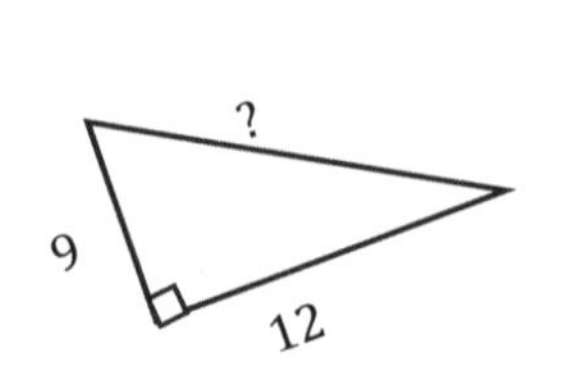

4)

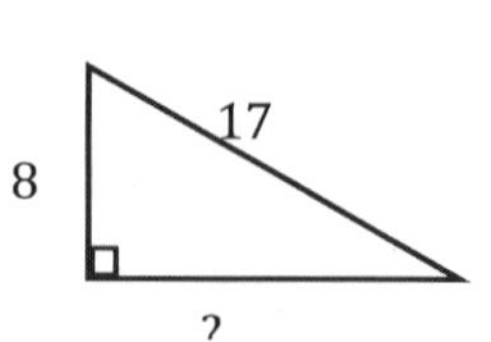

Complementary and Supplementary angles

- Two angles with a sum of 90 degrees are called complementary angles.

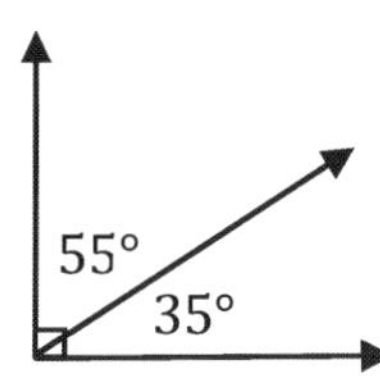

- Two angles with a sum of 180 degrees are Supplementary angles.

135° 45°

Examples:

Example 1. Angles Q and S are supplementary. What is the measure of angle Q if angle S is 35 degrees?

Solution: Q and S are supplementary → $Q + S = 180 \rightarrow Q + 35 = 180 \rightarrow$

$$Q = 180 - 35 = 145$$

Example 2. Angles x and y are complementary. What is the measure of angle x if angle y is 16 degrees?

Solution: Angles x and y are complementary → $x + y = 90 \rightarrow x + 16 = 90 \rightarrow$

$$x = 90 - 16 = 74$$

Practices:

✍ ***Find the missing measurement in the pair of angles.***

1) $x =$ ____

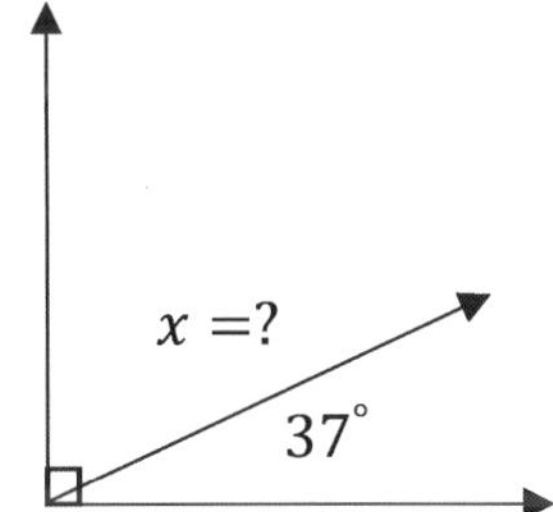

2) $x =$ ____

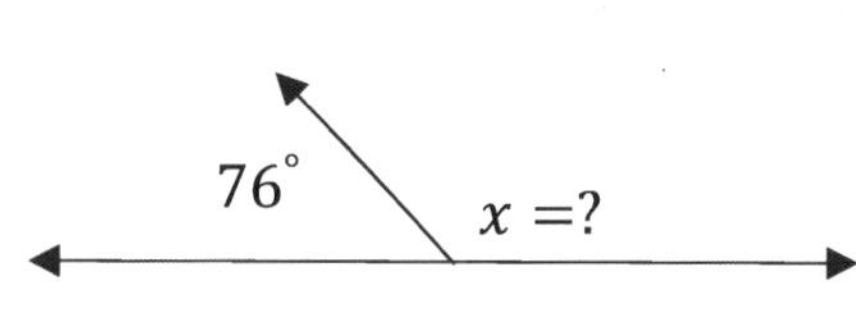

3) The measure of an angle is 49°. What is the measure of its complementary angle? ___

4) The measure of an angle is 143°. What is the measure of its supplementary angle?___

Parallel lines and Transversals

- When a line (transversal) intersects two parallel lines in the same plane, eight angles are formed. In the following diagram, a transversal intersects two parallel lines. Angles 1, 3, 5, and 7 are congruent. Angles 2, 4, 6, and 8 are also congruent.
- In the following diagram, the following angles are supplementary angles (their sum is 180):
  - Angles 1 and 8
  - Angles 2 and 7
  - Angles 3 and 6
  - Angles 4 and 5

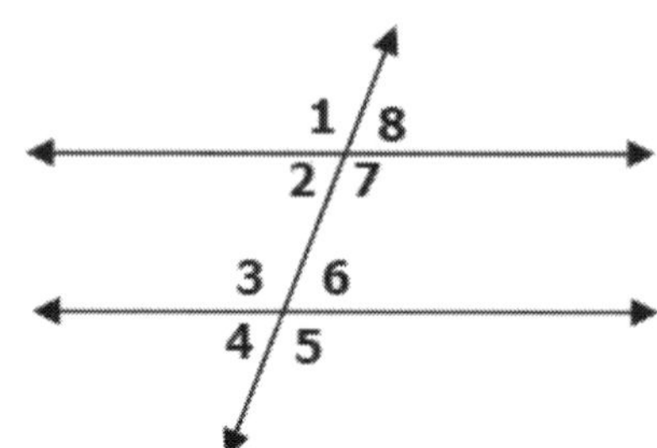

Example:

In the following diagram, two parallel lines are cut by a transversal. What is the value of x?

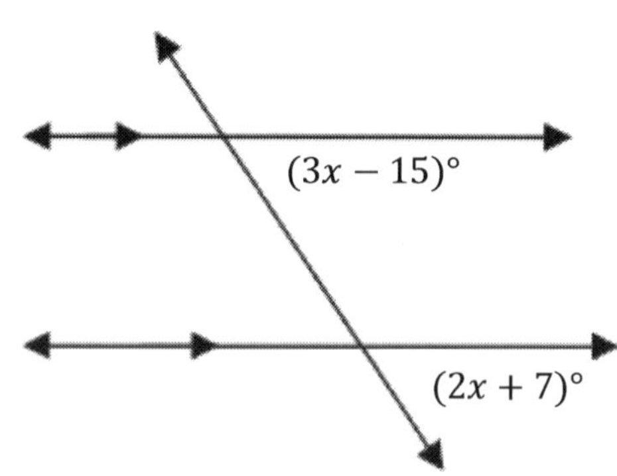

Solution: The two angles $3x - 15$ and $2x + 7$ are equivalent. That is: $3x - 15 = 2x + 7$

Now, solve for x: $3x - 15 + 15 = 2x + 7 + 15$

$\rightarrow 3x = 2x + 22 \rightarrow 3x - 2x = 2x + 22 - 2x \rightarrow$
$x = 22$

Practices:

In the following diagrams, solve for x.

1) $x =$ ____

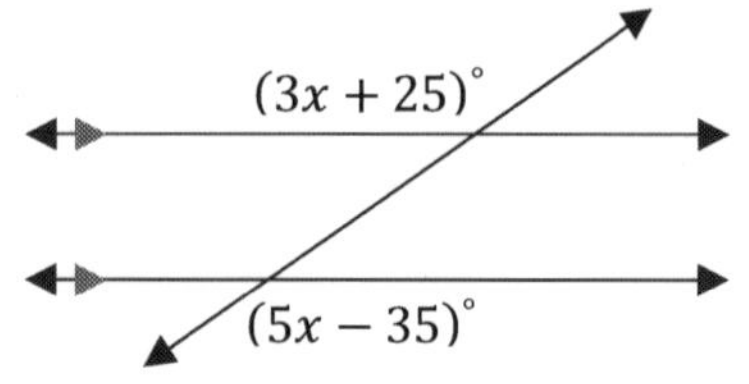

2) $x =$ ____

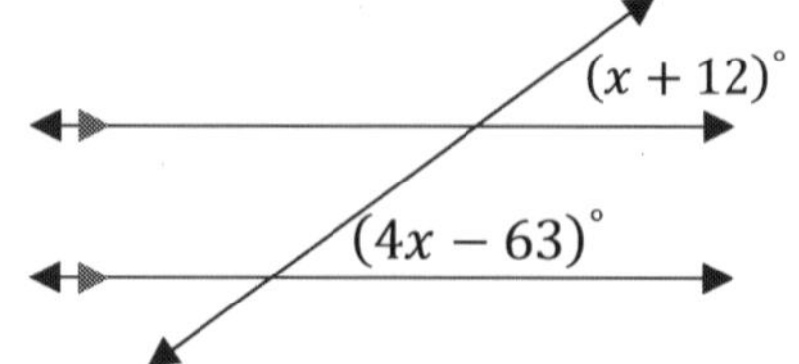

Triangles

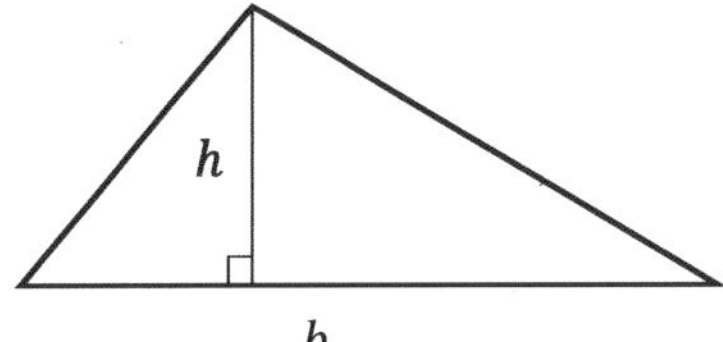

- In any triangle, the sum of all angles is 180 degrees.
- Area of a triangle $= \frac{1}{2}(base \times height)$

Examples:

Example 1. What is the area of this triangle?

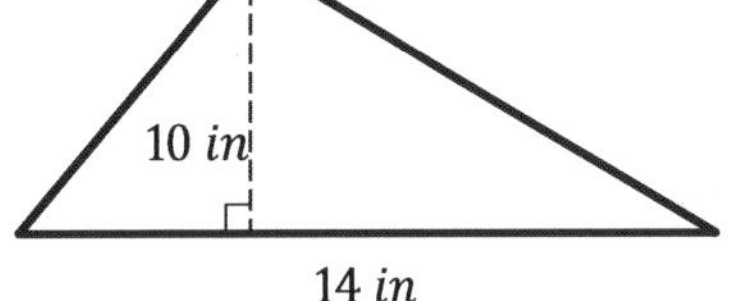

Solution: Use the area formula:

Area $= \frac{1}{2}(base \times height)$

$base = 14$ and $height = 10$, Then:

Area $= \frac{1}{2}(14 \times 10) = \frac{1}{2}(140) = 70\ in^2$

Example 2. What is the missing angle in this triangle?

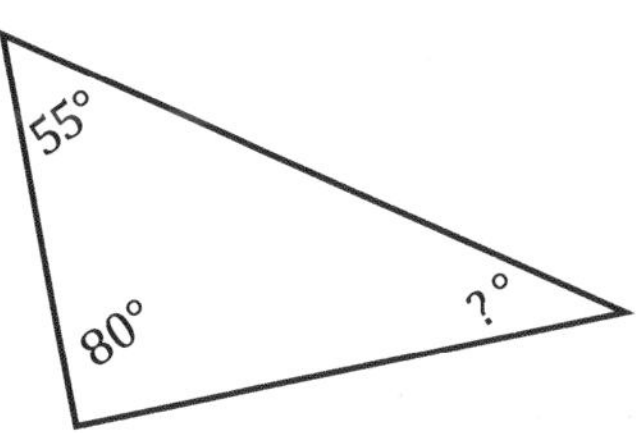

Solution: In any triangle, the sum of all angles is 180 degrees. Let x be the missing angle.

Then: $55 + 80 + x = 180 \rightarrow 135 + x = 180 \rightarrow$

$x = 180 - 135 = 45$

The missing angle is 45 degrees.

Practices:

Find the measure of the unknown angle in each triangle.

1) 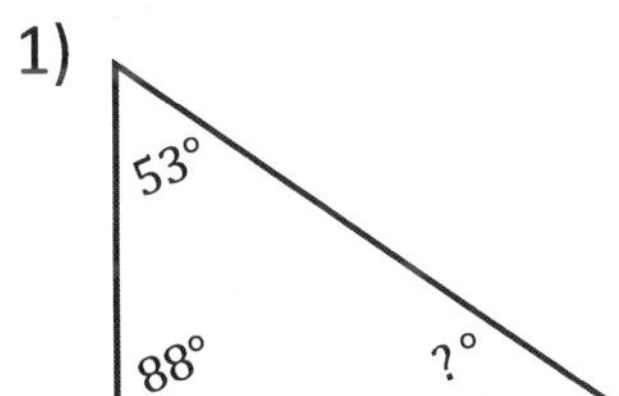

2) 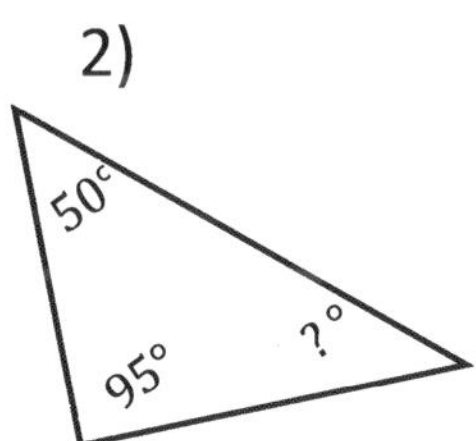

3) 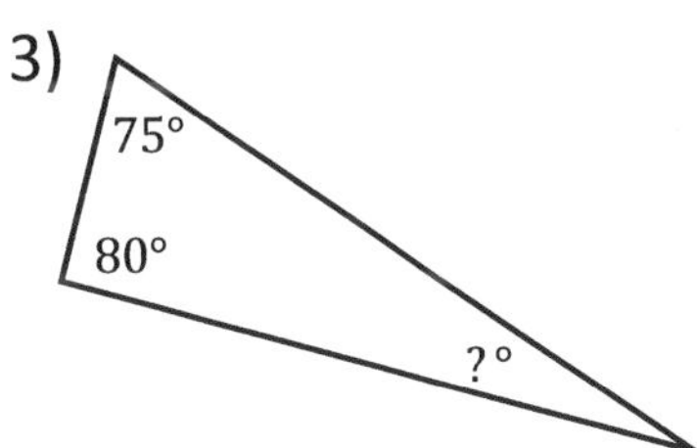

Find area of each triangle.

4) 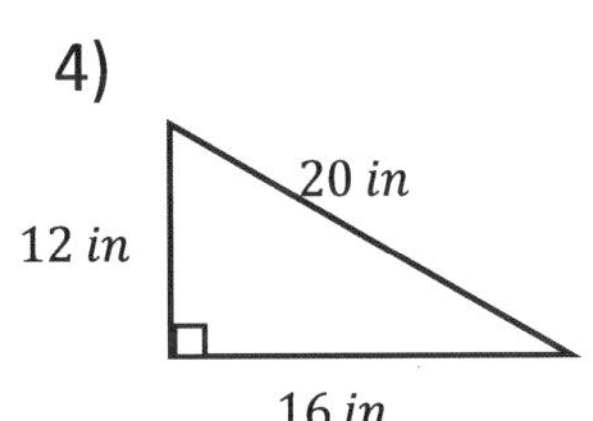

5) 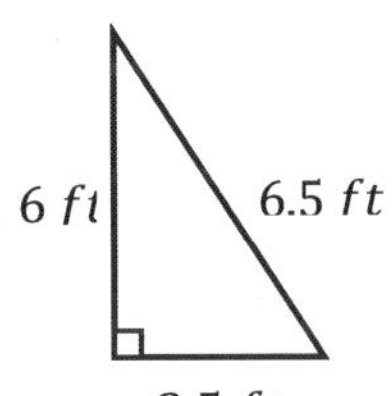

6) 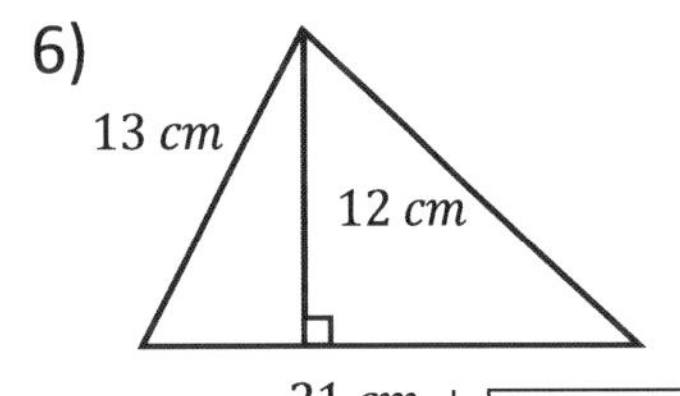

Find more at bit.ly/3haZrRg

Special Right Triangles

- A special right triangle is a triangle whose sides are in a particular ratio. Two special right triangles are $45^{\circ}-45^{\circ}-90^{\circ}$ and $30^{\circ}-60^{\circ}-90^{\circ}$ triangles.
- In a special $45^{\circ}-45^{\circ}-90^{\circ}$ triangle, the three angles are 45°, $45^{\circ}$and 90°. The lengths of the sides of this triangle are in the ratio of $1:1:\sqrt{2}$.
- In a special triangle $30^{\circ}-60^{\circ}-90^{\circ}$, the three angles are $30^{\circ}-60^{\circ}-90^{\circ}$. The lengths of this triangle are in the ratio of $1:\sqrt{3}:2$.

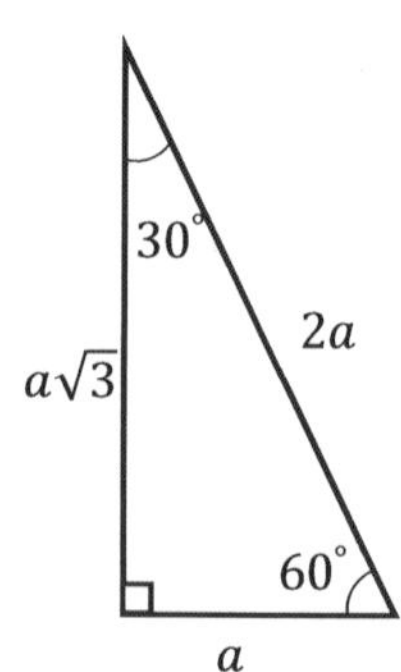

Examples:

Example 1. Find the length of the hypotenuse of a right triangle if the length of the other two sides are both 4 inches.

Solution: this is a right triangle with two equal sides. Therefore, it must be a $45^{\circ}-45^{\circ}-90^{\circ}$ triangle. Two equivalent sides are 4 inches. The ratio of sides: $x:x:x\sqrt{2}$

The length of the hypotenuse is $4\sqrt{2}$ inches. $x:x:x\sqrt{2} \rightarrow 4:4:4\sqrt{2}$

Example 2. The length of the hypotenuse of a right triangle is 6 inches. What are the lengths of the other two sides if one angle of the triangle is 30°?

Solution: The hypotenuse is 6 inches and the triangle is a $30^{\circ}-60^{\circ}-90^{\circ}$ triangle. Then, one side of the triangle is 3 (it's half the side of the hypotenuse) and the other side is $3\sqrt{3}$. (it's the smallest side times $\sqrt{3}$)

$x:x\sqrt{3}:2x \rightarrow x=3 \rightarrow x:x\sqrt{3}:2x=3:3\sqrt{3}:6$

Practices:

Find the value of x and y in each triangle.

1) $x=$ ___ $y=$ ___ 2) $x=$ ___ $y=$ ___ 3) $x=$ ___ $y=$ ___

x
15
y

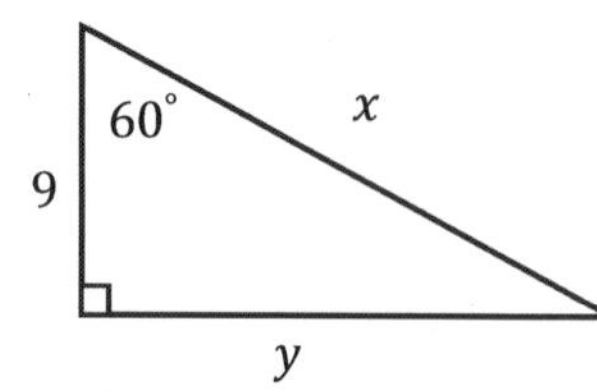

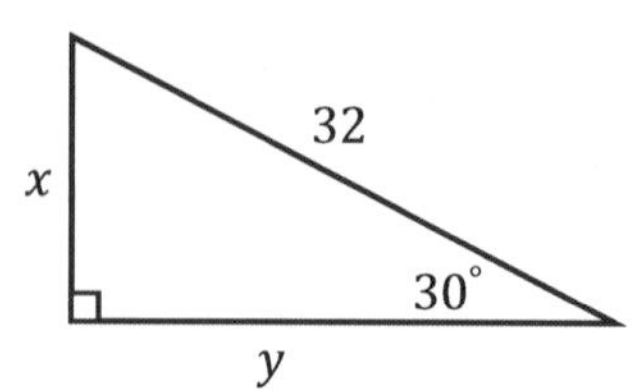

Polygons

- The perimeter of a square $= 4 \times side = 4s$

- The perimeter of a rectangle $= 2(width + length)$

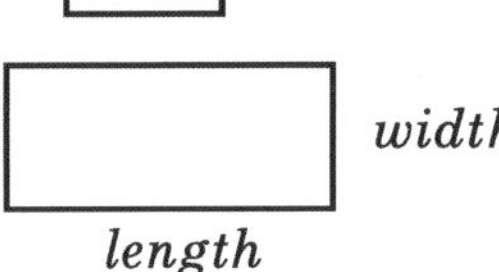

- The perimeter of a trapezoid $= a + b + c + d$

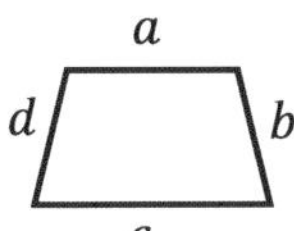

- The perimeter of a regular hexagon $= 6a$

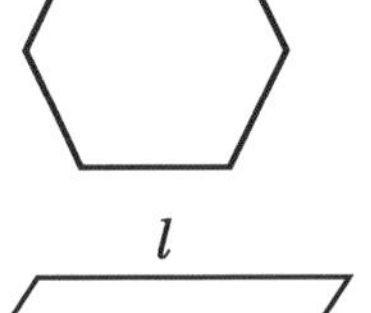

- The perimeter of a parallelogram $= 2(l + w)$

Examples:

Example 1. Find the perimeter of the following regular hexagon.

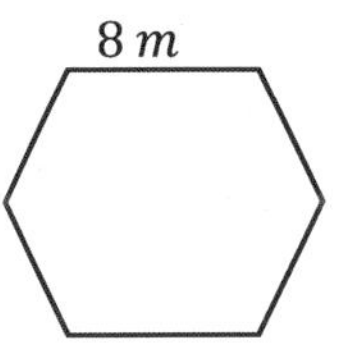

Solution: Since the hexagon is regular, all sides are equal.

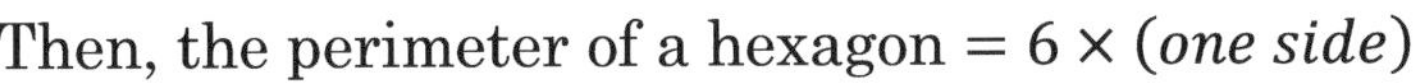
Then, the perimeter of a hexagon $= 6 \times (one\ side)$

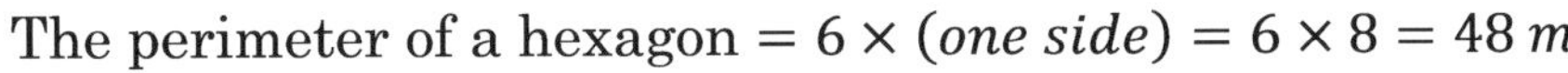
The perimeter of a hexagon $= 6 \times (one\ side) = 6 \times 8 = 48\ m$

Example 2. Find the perimeter of the following trapezoid.

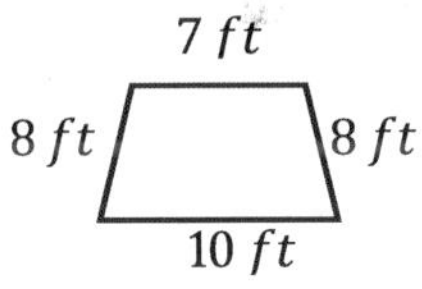

Solution: The perimeter of a trapezoid $= a + b + c + d$

The perimeter of a trapezoid $= 7 + 8 + 8 + 10 = 33\ ft$

Practices:

 ***Find the perimeter of each shape.***

1)

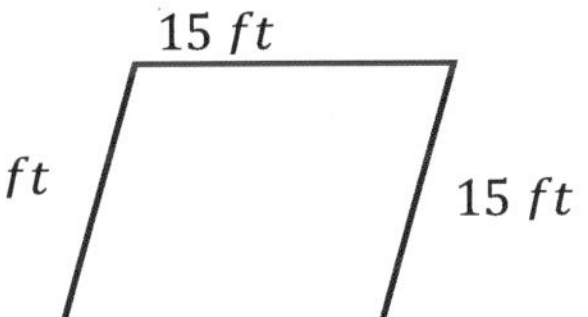

2)

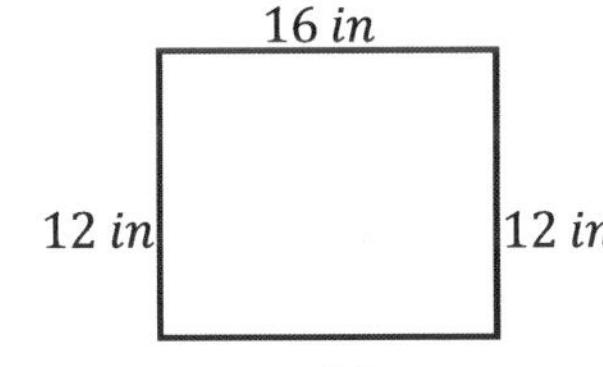

3) Regular hexagon

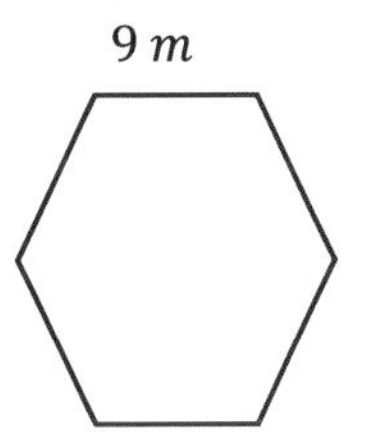

4) Square

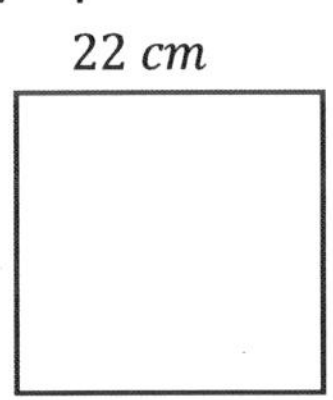

Circles

- In a circle, variable r is usually used for the radius and d for diameter.
- $Area\ of\ a\ circle = \pi r^2$ (π is about 3.14)
- $Circumference\ of\ a\ circle = 2\pi r$

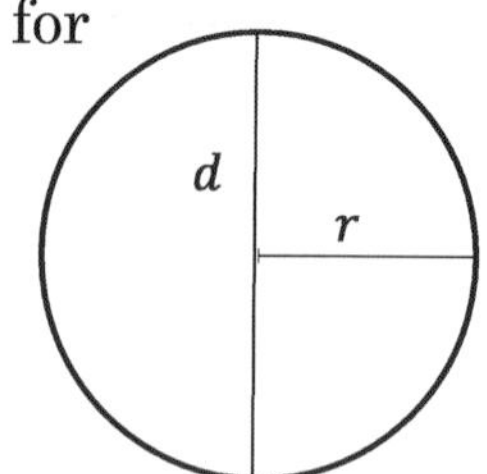

Examples:

Example 1. Find the area of this circle. ($\pi = 3.14$)

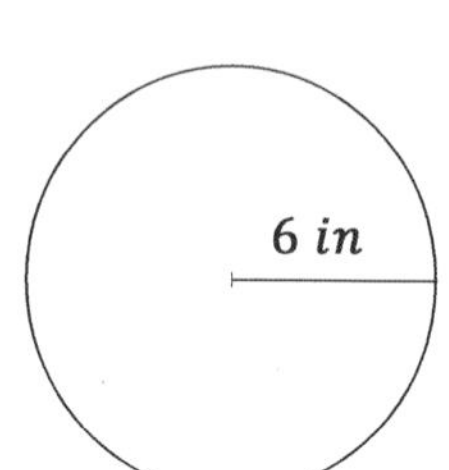

Solution:

Use area formula: $Area = \pi r^2$

$r = 6\ in \rightarrow Area = \pi(6)^2 = 36\pi$, ($\pi = 3.14$)

Then: $Area = 36 \times 3.14 = 113.04\ in^2$

Example 2. Find the Circumference of this circle. ($\pi = 3.14$)

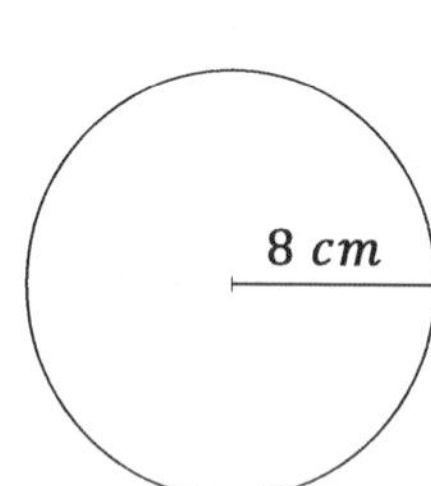

Solution:

Use Circumference formula: $Circumference = 2\pi r$

$r = 8\ cm \rightarrow Circumference = 2\pi(8) = 16\pi$

($\pi = 3.14$), Then: $Circumference = 16 \times 3.14 = 50.24\ cm$

Practices:

✍ ***Find the Circumference of each circle.*** ($\pi = 3.14$)

1) ____ 2) ____ 3) ____ 4) ____ 5) ____

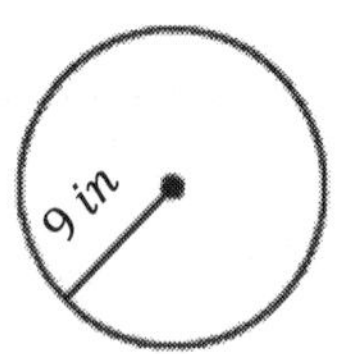

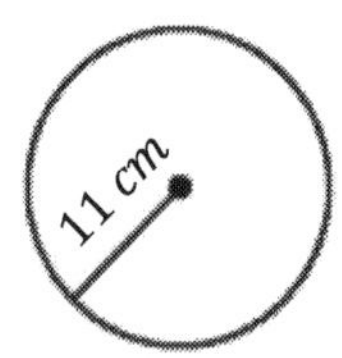

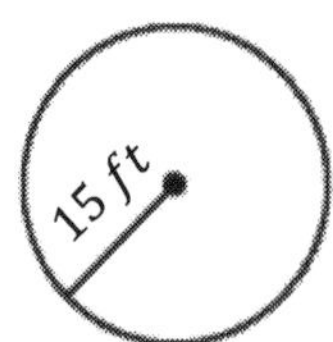

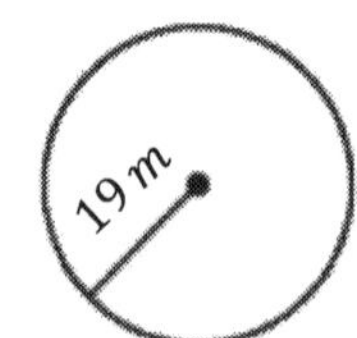

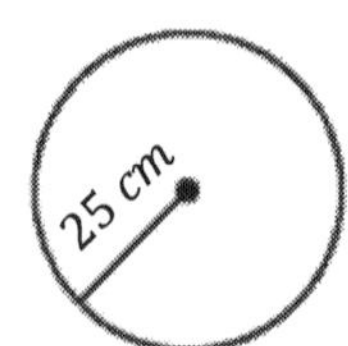

Trapezoids

- A quadrilateral with at least one pair of parallel sides is a trapezoid.
- Area of a trapezoid $= \frac{1}{2}h(b_1 + b_2)$

b_2

h

b_1

Examples:

Example 1. Calculate the area of this trapezoid.

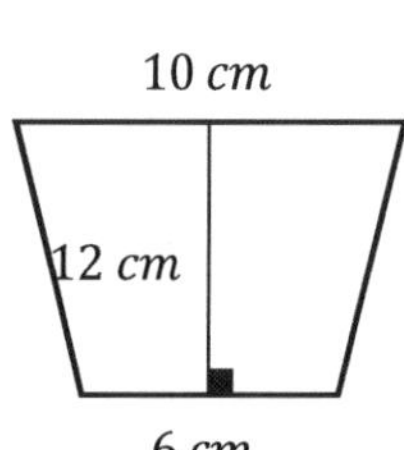

Solution:

Use area formula: $A = \frac{1}{2}h(b_1 + b_2)$

$b_1 = 6\ cm$, $b_2 = 10\ cm$ and $h = 12\ cm$

Then: $A = \frac{1}{2}(12)(10 + 6) = 6(16) = 96\ cm^2$

Example 2. Calculate the area of this trapezoid.

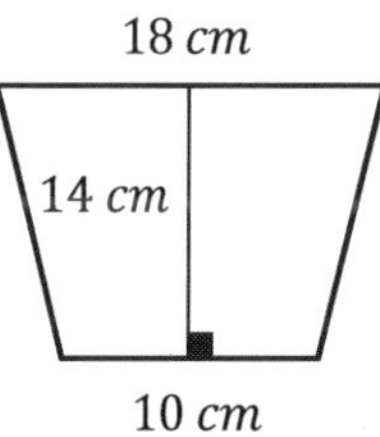

Solution:

Use area formula: $A = \frac{1}{2}h(b_1 + b_2)$

$b_1 = 10\ cm$, $b_2 = 18\ cm$ and $h = 14\ cm$

Then: $A = \frac{1}{2}(14)(10 + 18) = 196\ cm^2$

Practices:

Find the area of each trapezoid.

1)

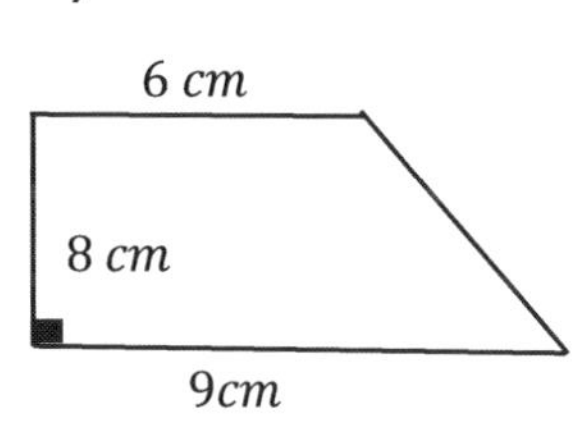

2)

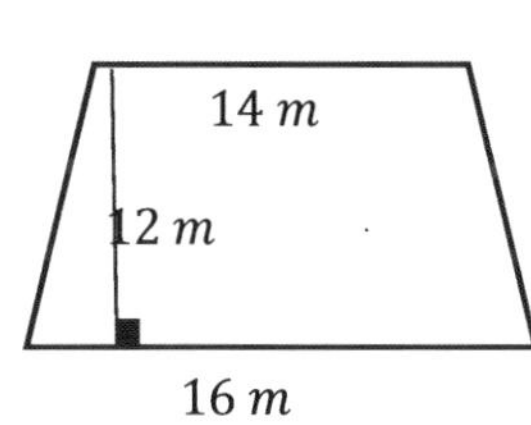

3)

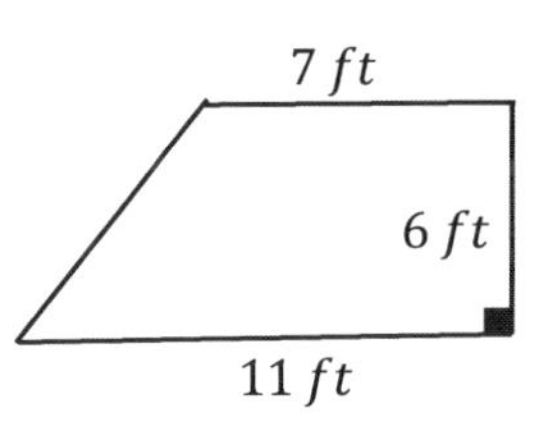

4)

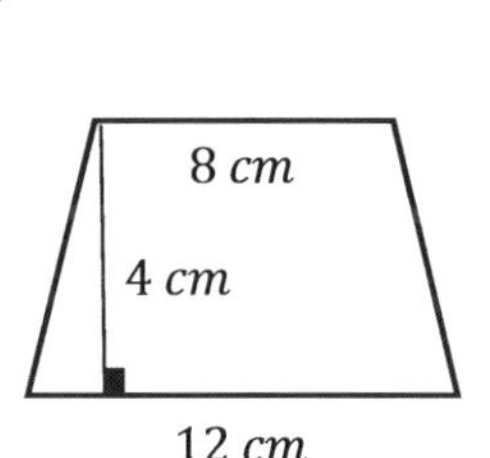

Cubes

- A cube is a three-dimensional solid object bounded by six square sides.
- Volume is the measure of the amount of space inside of a solid figure, like a cube, ball, cylinder, or pyramid.
- The volume of a cube $= (one\ side)^3$
- The surface area of a cube $= 6 \times (one\ side)^2$

Examples:

Example 1. Find the volume and surface area of this cube.

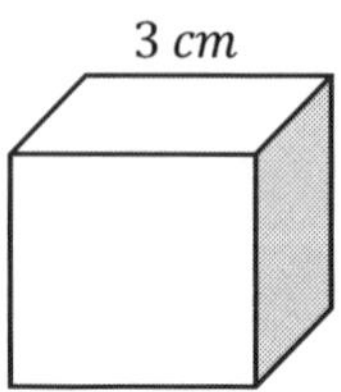

Solution: Use volume formula: $volume = (one\ side)^3$

Then: $volume = (one\ side)^3 = (3)^3 = 27\ cm^3$

Use surface area formula:

$surface\ area\ of\ a\ cube$: $6(one\ side)^2 = 6(3)^2 = 6(9) = 54\ cm^2$

Example 2. Find the volume and surface area of this cube.

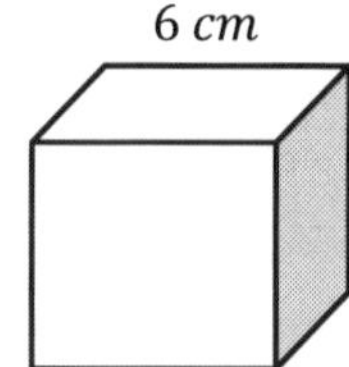

Solution: Use volume formula: $volume = (one\ side)^3$

Then: $volume = (one\ side)^3 = (6)^3 = 216\ cm^3$

Use surface area formula:

$surface\ area\ of\ a\ cube$: $6(one\ side)^2 = 6(6)^2 = 6(36) = 216\ cm^2$

Practices:

Find the volume of each cube

1) 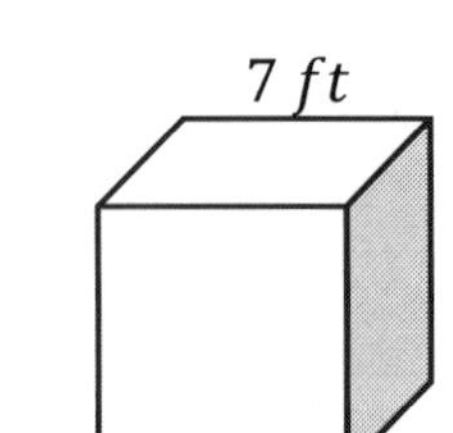

2) 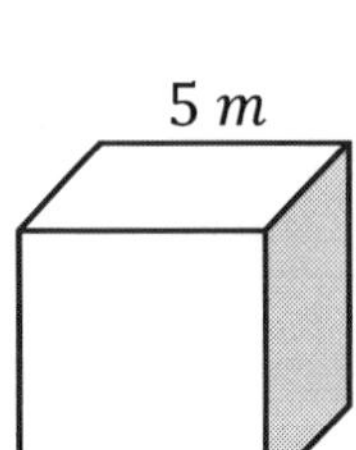

3) 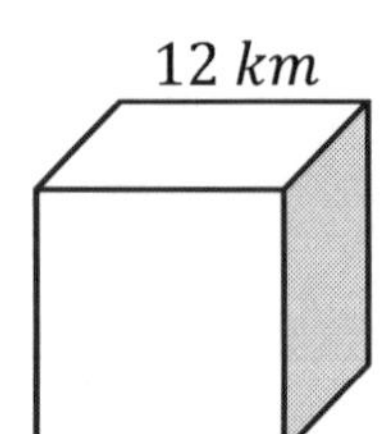

4) 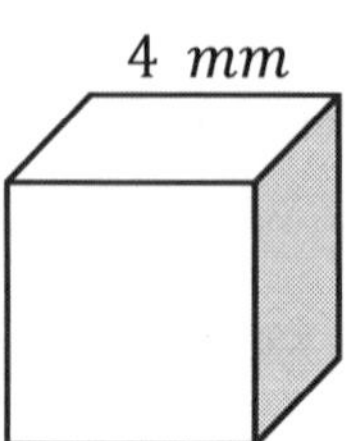

Rectangular Prisms

- A rectangular prism is a solid 3-dimensional object with six rectangular faces.
- The volume of a Rectangular prism = $Length \times Width \times Height$

  $Volume = l \times w \times h$

  $Surface\ area = 2 \times (wh + lw + lh)$

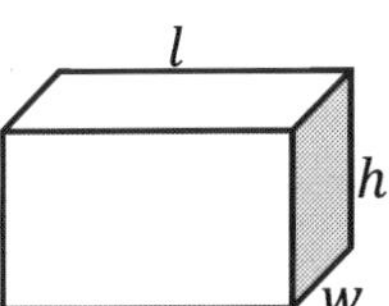

Examples:

Example 1. Find the volume and surface area of this rectangular prism.

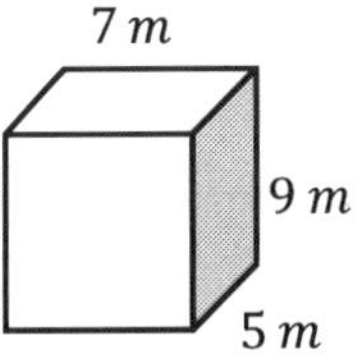

Solution: Use volume formula: $Volume = l \times w \times h$

Then: $Volume = 7 \times 5 \times 9 = 315\ m^3$

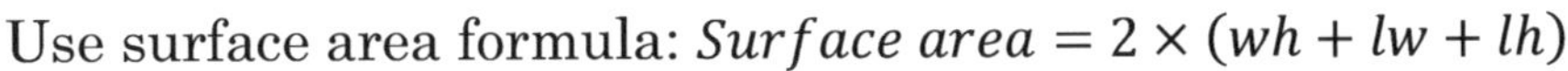

Use surface area formula: $Surface\ area = 2 \times (wh + lw + lh)$

Then: $Surface\ area = 2 \times \big((5 \times 9) + (7 \times 5) + (7 \times 9)\big)$

$= 2 \times (45 + 35 + 63) = 2 \times (143) = 286\ m^2$

Example 2. Find the volume and surface area of this rectangular prism.

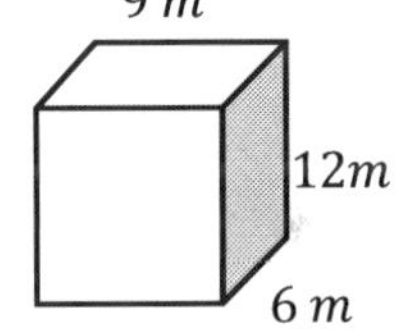

Solution: Use volume formula: $Volume = l \times w \times h$

Then: $Volume = 9 \times 6 \times 12 = 648\ m^3$

Use surface area formula: $Surface\ area = 2 \times (wh + lw + lh)$

Then: $Surface\ area = 2 \times \big((6 \times 12) + (9 \times 6) + (9 \times 12)\big)$

$= 2 \times (72 + 54 + 108) = 2 \times (234) = 468\ m^2$

Practices:

Find the volume of each Rectangular Prism.

1) 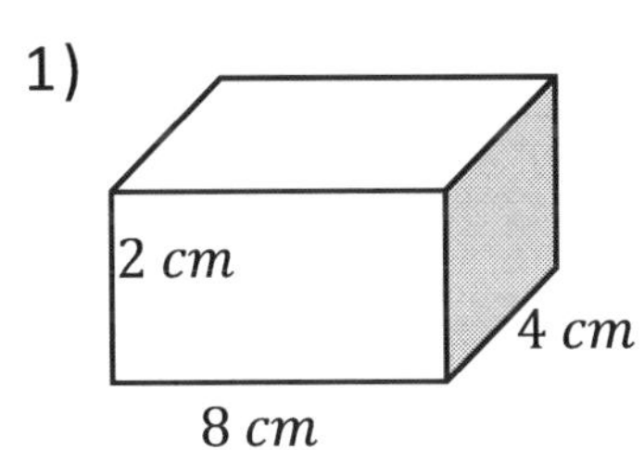

2) 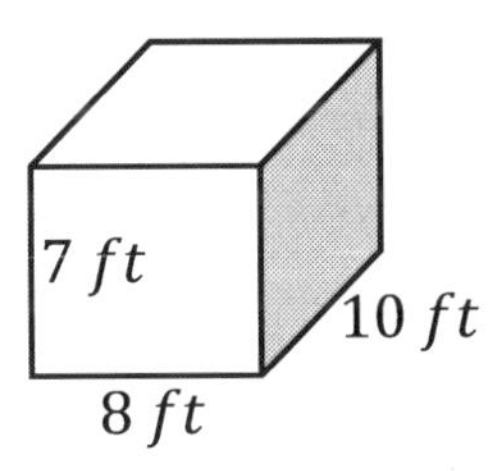

3) 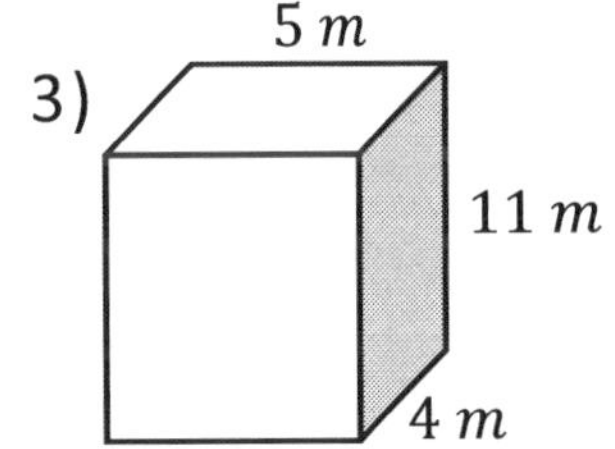

bit.ly/3nKm2GT
Find more at

Cylinder

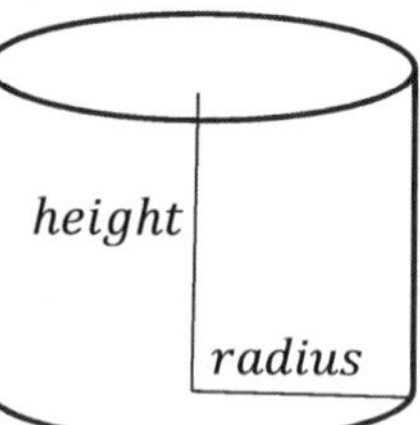

- A cylinder is a solid geometric figure with straight parallel sides and a circular or oval cross-section.
- $The\ volume\ of$ a $cylinder = \pi(radius)^2 \times height,\ \pi \approx 3.14$
- $The\ surface\ area\ of\ a\ cylinder = 2\pi r^2 + 2\pi rh$

Examples:

Example 1. Find the volume and Surface area of the following Cylinder.

Solution: Use volume formula: $Volume = \pi(radius)^2 \times height.$

Then: $Volume = \pi(4)^2 \times 10 = 16\pi \times 10 = 160\pi$

$\pi = 3.14$, then: $Volume = 160\pi = 160 \times 3.14 = 502.4\ cm^3$

Use surface area formula: $Surface\ area = 2\pi r^2 + 2\pi rh$

Then: $2\pi(4)^2 + 2\pi(4)(10) = 2\pi(16) + 2\pi(40) = 32\pi + 80\pi = 112\pi$

$\pi = 3.14$, Then: $Surface\ area = 112 \times 3.14 = 351.68\ cm^2$

Example 2. Find the volume and Surface area of the following Cylinder.

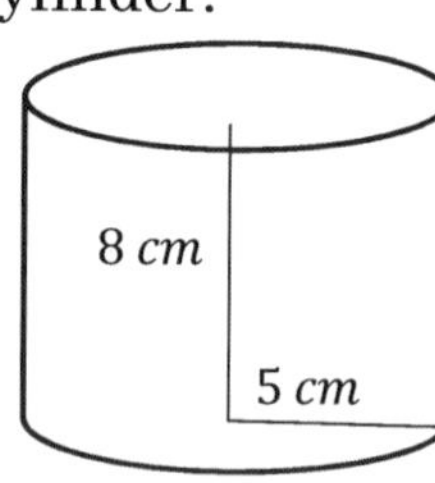

Solution: Use volume formula: $Volume = \pi(radius)^2 \times height.$

Then: $Volume = \pi(5)^2 \times 8 = 25\pi \times 8 = 200\pi$

$\pi = 3.14$, Then: $Volume = 200\pi = 628\ cm^3$

Use surface area formula: $Surface\ area = 2\pi r^2 + 2\pi rh$

Then: $2\pi(5)^2 + 2\pi(5)(8) = 2\pi(25) + 2\pi(40) = 50\pi + 80\pi = 130\pi$

$\pi = 3.14$ then: $Surface\ area = 130 \times 3.14 = 408.2\ cm^2$

Practices:

✍ ***Find the volume of each Cylinder. Round your answer to the nearest tenth.*** $(\pi = 3.14)$

1) 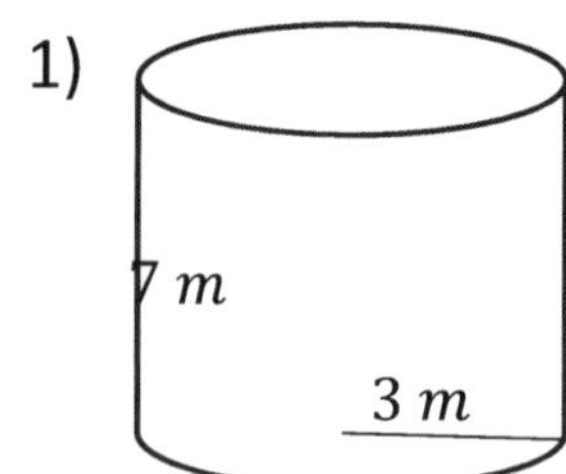

2) 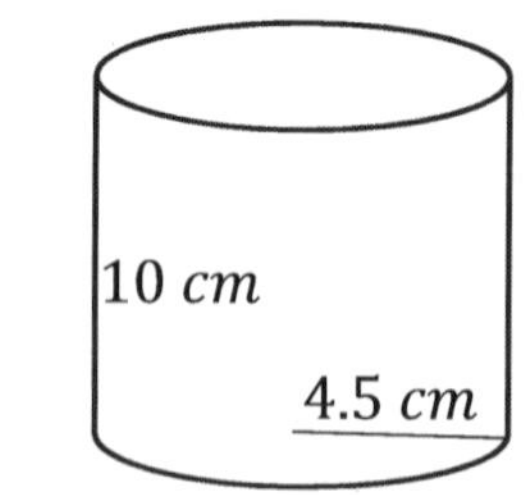

3) 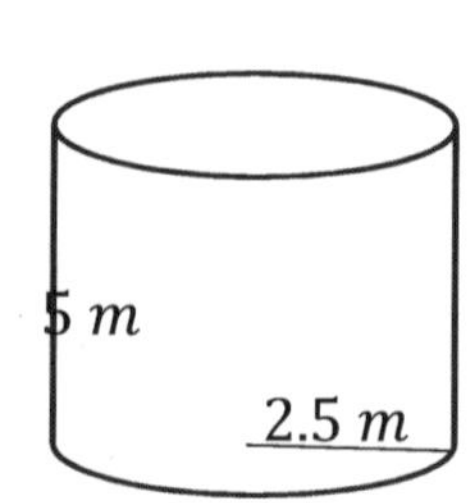

Chapter 11: Answers

The Pythagorean Theorem

1) 13 (Use Pythagorean Theorem: $a^2 + b^2 = c^2$. Then: $a^2 + b^2 = c^2 \rightarrow 5^2 + 12^2 = c^2 \rightarrow 25 + 144 = c^2 \rightarrow c^2 = 169 \rightarrow c = \sqrt{169} = 13$)

2) 20 (Use Pythagorean Theorem: $a^2 + b^2 = c^2$. Then: $a^2 + b^2 = c^2 \rightarrow 16^2 + 12^2 = c^2 \rightarrow 256 + 144 = c^2 \rightarrow c^2 = 400 \rightarrow c = \sqrt{400} = 20$)

3) 15 (Use Pythagorean Theorem: $a^2 + b^2 = c^2$. Then: $a^2 + b^2 = c^2 \rightarrow 9^2 + 12^2 = c^2 \rightarrow 81 + 144 = c^2 \rightarrow c^2 = 225 \rightarrow c = \sqrt{225} = 15$)

4) 15 (Use Pythagorean Theorem: $a^2 + b^2 = c^2$. Then: $a^2 + b^2 = c^2 \rightarrow 8^2 + b^2 = 17^2 \rightarrow 64 + b^2 = 289 \rightarrow b^2 = 289 - 64 = 225 \rightarrow b = \sqrt{225} = 15$)

Complementary and Supplementary Angles

1) $x = 53°$ (Notice that the two angles form a right angle. This means that the angles are complementary, and their sum is 90. Then: $37^\circ + x = 90^\circ \rightarrow x = 90^\circ - 37^\circ = 53^\circ$. The missing angle is 53 degrees. $x = 53°$)

2) $x = 104°$ (The angles are supplementary, and their sum is 180. Then: $76^\circ + x = 180 \rightarrow x = 180^\circ - 76^\circ = 104^\circ$. The missing angle is 104 degrees. $x = 104°$)

3) $x = 41°$ ($49^\circ + x = 90^\circ \rightarrow x = 90^\circ - 49^\circ = 41^\circ$)

4) $x = 37°$ ($143^\circ + x = 180^\circ \rightarrow x = 180^\circ - 143^\circ = 37^\circ$)

Parallel Lines and Transversals

1) $x = 30$ (The two angles $3x + 25$ and $5x - 35$ are equivalent. That is: $3x + 25 = 5x - 35$. Now, solve for x : $3x + 25 - 25 = 5x - 35 - 25 \rightarrow 3x = 5x - 60 \rightarrow 3x - 5x = 5x - 60 - 5x \rightarrow 2x = 60 \rightarrow x = 30$)

2) $x = 25$ (The two angles $x + 12$ and $4x - 63$ are equivalent. That is: $x + 12 = 4x - 63$. Now, solve for x: $x + 12 - 12 = 4x - 63 - 12 \rightarrow x = 4x - 75 \rightarrow x - 4x = 4x - 75 - 4x \rightarrow -3x = -75 \rightarrow \frac{-3x}{-3} = \frac{-75}{-3} \rightarrow x = 25$)

Triangles

1) 39 (In any triangle, the sum of all angles is 180 degrees. Let x be the missing angle. Then: $53 + 88 + x = 180 \rightarrow 141 + x = 180 \rightarrow x = 180 - 141 = 39$)

2) 35 (In any triangle, the sum of all angles is 180 degrees. Let x be the missing angle. Then: $50 + 95 + x = 180 \rightarrow 145 + x = 180 \rightarrow x = 180 - 145 = 35$)

3) 25 (In any triangle, the sum of all angles is 180 degrees. Let x be the missing angle. Then: $80 + 75 + x = 180 \rightarrow 155 + x = 180 \rightarrow x = 180 - 155 = 25$)

4) $96\ in^2$ (Use the area formula: Area $= \frac{1}{2}(base \times height)$. $base = 16$ and $height = 12$, Then: Area $= \frac{1}{2}(16 \times 12) = \frac{1}{2}(192) = 96\ in^2$)

5) $7.5\ ft^2$ (Use the area formula: Area $= \frac{1}{2}(base \times height)$. $base = 2.5$ and $height = 6$, Then: Area $= \frac{1}{2}(2.5 \times 6) = \frac{1}{2}(15) = 7.5\ ft^2$)

6) $126\ cm^2$ (Use the area formula: Area $= \frac{1}{2}(base \times height)$. $base = 21$ and $height = 12$, Then: Area $= \frac{1}{2}(21 \times 12) = \frac{1}{2}(252) = 126\ cm^2$)

Special Right Triangles

1) $x = 15\sqrt{2}, y = 15$ (This shape is a square so all its sides are equal and the two triangles are $45^\circ - 45^\circ - 90^\circ$ triangle. Two equivalent sides are 15. The ratio of sides: $x : x : x\sqrt{2}$. The length of the hypotenuse is $15\sqrt{2}$. $x = 15\sqrt{2}$)

2) $x = 18, y = 9\sqrt{3}$ (The smallest side is 9 inches and the triangle is a $30^\circ - 60^\circ - 90^\circ$ triangle. Then, one side of the triangle is 18 (Hypotenuse is twice the small side) and the other side is $9\sqrt{3}$. $9 : 9\sqrt{3} : 18 \rightarrow x = 18$, $y = 9\sqrt{3}$)

3) $x = 16, y = 16\sqrt{3}$ (The hypotenuse is 32 and the triangle is a $30^\circ - 60^\circ - 90^\circ$ triangle. Then, one side of the triangle is 16 (it's half the side of the hypotenuse) and the other side is $16\sqrt{3}$. (it's the smallest side times $\sqrt{3}$) $16: 16\sqrt{3}: 32 \rightarrow x = 16, y = 16\sqrt{3}$)

Polygons

1) $60\ ft$ (Since all the sides of the shape are equal, the perimeter is equal to: $4 \times (one\ side) = 4 \times 15 = 60\ ft$)
2) $56\ in$ (The perimeter of a rectangle $= 2(width + length)$. The perimeter of the rectangle $= 2(12 + 16) = 56\ in$)
3) $54\ m$ (Since the hexagon is regular, all sides are equal. Then, the perimeter of the hexagon $= 6 \times (one\ side)$. The perimeter of the hexagon $= 6 \times (one\ side) = 6 \times 9 = 54\ m$)
4) $88\ cm$ (The perimeter of a square $= 4 \times side = 4s$. The perimeter of the square $= 4 \times 22 = 88\ cm$)

Circles

1) $56.52\ in$ (Use Circumference formula: $Circumference = 2\pi r$. $r = 9\ in \rightarrow Circumference = 2\pi(9) = 2 \times 3.14 \times 9 = 56.52\ in$)
2) $69.08\ cm$ (Use Circumference formula: $Circumference = 2\pi r$. $r = 11\ cm \rightarrow Circumference = 2\pi(11) = 2 \times 3.14 \times 11 = 69.08\ cm$)
3) $94.2\ ft$ (Use Circumference formula: $Circumference = 2\pi r$. $r = 15\ ft \rightarrow Circumference = 2\pi(15) = 2 \times 3.14 \times 15 = 94.2\ ft$)
4) $119.32\ m$ (Use Circumference formula: $Circumference = 2\pi r$. $r = 19\ m \rightarrow Circumference = 2\pi(19) = 2 \times 3.14 \times 19 = 119.32\ m$)
5) $157\ cm$ (Use Circumference formula: $Circumference = 2\pi r$. $r = 25\ cm \rightarrow Circumference = 2\pi(25) = 2 \times 3.14 \times 25 = 157\ cm$)

Trapezoids

1) $60\ cm^2$ (Use area formula: $A = \frac{1}{2}h(b_1 + b_2) \rightarrow b_1 = 6\ cm, b_2 = 9\ cm$ and $h = 8\ cm$. Then: $A = \frac{1}{2}(8)(6 + 9) = 4(15) = 60\ cm^2$)
2) $180\ m^2$ (Use area formula: $A = \frac{1}{2}h(b_1 + b_2) \rightarrow b_1 = 14\ cm, b_2 = 16\ cm$ and $h = 12\ cm$. Then: $A = \frac{1}{2}(12)(14 + 16) = 6(30) = 180\ m^2$)
3) $54\ ft^2$ (Use area formula: $A = \frac{1}{2}h(b_1 + b_2) \rightarrow b_1 = 7\ cm, b_2 = 11\ cm$ and $h = 6\ cm$. Then: $A = \frac{1}{2}(6)(7 + 11) = 3(18) = 54\ ft^2$)
4) $40\ cm^2$ (Use area formula: $A = \frac{1}{2}h(b_1 + b_2) \rightarrow b_1 = 8\ cm, b_2 = 12\ cm$ and $h = 4\ cm$. Then: $A = \frac{1}{2}(4)(8 + 12) = 2(20) = 40\ cm^2$)

Cubes

1) $343\ ft^3$(Use volume formula: $volume = (one\ side)^3$. Then: $volume = (one\ side)^3 = (7)^3 = 343\ ft^3$)
2) $125\ m^3$ (Use volume formula: $volume = (one\ side)^3$. Then: $volume = (one\ side)^3 = (5)^3 = 125\ m^3$)
3) $1{,}728\ km^3$ (Use volume formula: $volume = (one\ side)^3$. Then: $volume = (one\ side)^3 = (12)^3 = 1{,}728 km^3$)
4) $64\ mm^3$ (Use volume formula: $volume = (one\ side)^3$. Then: $volume = (one\ side)^3 = (42)^3 = 64\ mm^3$)

Rectangular Prisms

1) $64\ cm^3$(Use volume formula: $Volume = l \times w \times h$. Then: $Volume = 8 \times 4 \times 2 = 64\ cm^3$)

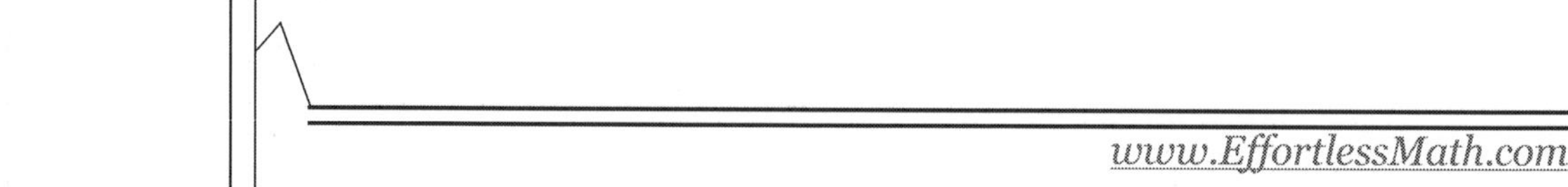

2) $560\ ft^3$ (Use volume formula: $Volume = l \times w \times h$. Then:
$Volume = 8 \times 10 \times 7 = 560\ ft^3$)

3) $220\ m^3$ (Use volume formula: $Volume = l \times w \times h$. Then:
$Volume = 5 \times 4 \times 11 = 220\ m^3$)

Cylinder

1) $197.8\ m^3$(Use volume formula: $Volume = \pi(radius)^2 \times height$. Then:
$Volume = \pi(3)^2 \times 7 = 9\pi \times 7 = 63\pi \rightarrow \pi = 3.14$, then:
$Volume = 63\pi = 63 \times 3.14 = 197.82 \approx 197.8\ m^3$)

2) $635.9\ cm^3$ (Use volume formula: $Volume = \pi(radius)^2 \times height$. Then:
$Volume = \pi(4.5)^2 \times 10 = 20.25\pi \times 10 = 202.5\pi \rightarrow \pi = 3.14$, then:
$Volume = 202.5\pi = 202.5 \times 3.14 = 635.85\ cm^3 \approx 635.9\ cm^3$)

3) $98.1\ m^3$ (Use volume formula: $Volume = \pi(radius)^2 \times height$. Then:
$Volume = \pi(2.5)^2 \times 5 = 6.25\pi \times 5 = 31.25\pi \rightarrow \pi = 3.14$, then:
$Volume = 31.25\pi = 31.25 \times 3.14 = 98.125\ m^3 \approx 98.1\ m^3$)

CHAPTER

12 Statistics

Math topics that you'll learn in this chapter:

☑ Mean, Median, Mode, and Range of the Given Data

☑ Pie Graph

☑ Probability Problems

☑ Permutations and Combinations

Mean, Median, Mode, and Range of the Given Data

- **Mean:** $\frac{sum\ of\ the\ data}{total\ number\ of\ data\ entires}$
- **Mode:** the value in the list that appears most often
- **Median:** is the middle number of a group of numbers arranged in order by size.
- **Range:** the difference between the largest value and smallest value in the list

Examples:

Example 1. What is the median of these numbers? $6, 11, 15, 10, 17, 20, 7$

Solution: Write the numbers in order: $6, 7, 10, 11, 15, 17, 20$
The median is the number in the middle. Therefore, the median is 11.

Example 2. What is the mean of these numbers? $7, 2, 3, 2, 4, 8, 7, 5$

Solution: Mean: $\frac{sum\ of\ the\ data}{total\ number\ of\ data\ entires} = \frac{7+2+3+2+4+8+7+5}{8} = \frac{38}{8} = 4.75$

Practices:

Solve.

1) In a javelin throw competition, five athletics score 75, 78, 80, 76 and 84 meters. What are their Mean and Median? ________________

2) Eva went to shop and bought 5 apples, 3 peaches, 11 bananas, 6 pineapple and 2 melons. What are the Mean and Median of her purchase? __________

Find Mode and Rage of the Given Data.

3) $12, 8, 11, 10, 2, 5, 11$

Mode: _____ Range: _____

4) $52, 53, 45, 50, 53, 54, 52, 53, 56$

Mode: _____ Range: _____

5) $6, 5, 8, 10, 6, 9, 6, 5, 8, 10$

Mode: _____ Range: _____

6) $17, 10, 9, 9, 3, 3, 9, 11$

Mode: _____ Range: _____

Pie Graph

- A Pie Graph (Pie Chart) is a circle chart divided into sectors, each sector represents the relative size of each value.
- Pie charts represent a snapshot of how a group is broken down into smaller pieces.

Example:

A library has 750 books that include Mathematics, Physics, Chemistry, English and History. Use the following graph to answer the questions.

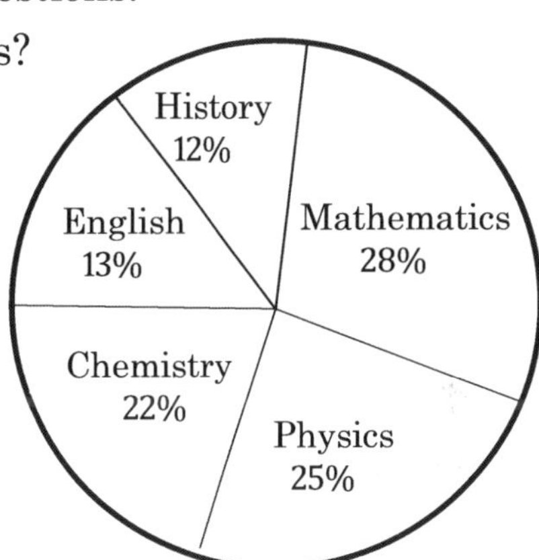

Example 1. What is the number of Mathematics books?

Solution: Number of total books = 750

Percent of Mathematics books = 28%

Then, the number of Mathematics books: 28% × 750 = 0.28 × 750 = 210

Example 2. What is the number of History books?

Solution: Number of total books = 750

Percent of History books = 12%

Then: 0.12 × 750 = 90

Practices:

The circle graph below shows all jane's expenses for last month. jane spent* $300 *on his bills last month.

1) How much did jane spend on his car last month? ________
2) How much did jane spend for foods last month? ________
3) How much did jane spend on his rent last month? ______
4) What fraction is jane's expenses for his bills and Car out of his total expenses last month?

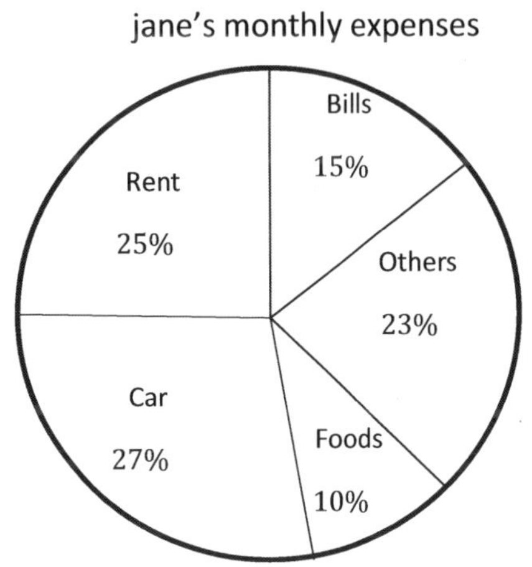

Probability Problems

- Probability is the likelihood of something happening in the future. It is expressed as a number between zero (can never happen) to 1 (will always happen).
- Probability can be expressed as a fraction, a decimal, or a percent.
- Probability formula: $Probability = \frac{number\ of\ desired\ outcomes}{number\ of\ total\ outcomes}$

Examples:

Example 1. Anita's trick–or–treat bag contains 10 pieces of chocolate, 16 suckers, 16 pieces of gum, 22 pieces of licorice. If she randomly pulls a piece of candy from her bag, what is the probability of her pulling out a piece of sucker?

Solution: $Probability = \frac{number\ of\ desired\ outcomes}{number\ of\ total\ outcomes}$

$Probability\ of\ pulling\ out\ a\ piece\ of\ sucker = \frac{16}{10+16+16+22} = \frac{16}{64} = \frac{1}{4}$

Example 2. A bag contains 20 balls: four green, five black, eight blue, a brown, a red and one white. If 19 balls are removed from the bag at random, what is the probability that a brown ball has been removed?

Solution: If 19 balls are removed from the bag at random, there will be one ball in the bag. The probability of choosing a brown ball is 1 out of 20. Therefore, the probability of not choosing a brown ball is 19 out of 20 and the probability of having not a brown ball after removing 19 balls is the same. The answer is: $\frac{19}{20}$

Practices:

Solve.

1) A number is chosen at random from 1 to 15. Find the probability of selecting an even number. ____________

2) In one bag there are 4 red marbles, 2 yellow marbles and 3 white marbles. We accidentally pull out a nut. What is the probability of red marbles coming out? What is the probability that white marbles will not come out? ____________

Permutations and Combinations

- **Factorials** are products, indicated by an exclamation mark. For example, $4! = 4 \times 3 \times 2 \times 1$ (Remember that 0! is defined to be equal to 1)
- **Permutations:** The number of ways to choose a sample of k elements from a set of n distinct objects where order does matter, and replacements are not allowed. For a permutation problem, use this formula:
$$_{n}P_{k} = \frac{n!}{(n-k)!}$$
- **Combination:** The number of ways to choose a sample of r elements from a set of n distinct objects where order does not matter, and replacements are not allowed. For a combination problem, use this formula:
$$_{n}C_{r} = \frac{n!}{r!\,(n-r)!}$$

Examples:

Example 1. How many ways can the first and second place be awarded to 7 people?

Solution: Since the order matters, (the first and second place are different!) we need to use permutation formula where n is 7 and k is 2. Then: $\frac{n!}{(n-k)!} = \frac{7!}{(7-2)!} = \frac{7!}{5!} = \frac{7\times6\times5!}{5!}$, remove 5! from both sides of the fraction. Then: $\frac{7\times6\times5!}{5!} = 7 \times 6 = 42$

Example 2. How many ways can we pick a team of 3 people from a group of 8?

Solution: Since the order doesn't matter, we need to use a combination formula where n is 8 and r is 3.

Then: $\frac{n!}{r!\,(n-r)!} = \frac{8!}{3!\,(8-3)!} = \frac{8!}{3!\,(5)!} = \frac{8\times7\times6\times5!}{3!\,(5)!} = \frac{8\times7\times6}{3\times2\times1} = \frac{336}{6} = 56$

Practices:

Solve.

1) How can 4 people be selected from 6 students to participate in a mountaineering team? ____________
2) How many ways can you give 6 flowers to your 10 friends? ____________
3) In a competition out of 14 participants, how many ways can three gold, silver and bronze medals reach the participants? ____________
4) In how many ways can a teacher chooses 13 out of 16 students? ____________

Chapter 12: Answers

Mean, Median, Mode, and Range of the Given Data

1) Mean: 78.6 median:78 (Write the numbers in order: $75, 76, 78, 80, 84$. The median is the number in the middle. Therefore, the median is 78.
Mean: $\frac{sum\ of\ the\ data}{total\ number\ of\ data\ entires} = \frac{75+78+80+76+84}{5} = \frac{393}{5} = 78.6$)

2) Mean: 5.4 median: 5 (Write the numbers in order: $2, 3, 5, 6, 11$. The median is the number in the middle. Therefore, the median is 5.
Mean: $\frac{sum\ of\ the\ data}{total\ number\ of\ data\ entires} = \frac{2+3+5+6+11}{5} = \frac{27}{5} = 5.4$)

3) Mode: 11 Range: 10 (The mode is number 11. There are two number 11 in the data. Range is the difference of the largest value and smallest value in the list. The largest value is 12 and the smallest value is 2. Then: $12 - 2 = 10$)

4) Mode: 53 Range: 11 (The mode is number 53. There are three number 53 in the data. Range is the difference of the largest value and smallest value in the list. The largest value is 56 and the smallest value is 45. Then: $56 - 45 = 11$)

5) Mode: 6 Range: 5 (The mode is number 6. There are three number 6 in the data. Range is the difference of the largest value and smallest value in the list. The largest value is 10 and the smallest value is 5. Then: $10 - 5 = 5$)

6) Mode: 9 Range: 14 (The mode is number 9. There are three number 9 in the data. Range is the difference of the largest value and smallest value in the list. The largest value is 17 and the smallest value is 3. Then: $17 - 3 = 14$)

Pie Graph

1) 540 (jane spent on his bills $= 15\% \times$ the total amount of jane's expenses$\rightarrow$ $300 = 15\% \times$the total amount of jane's expense$\rightarrow$ the total amount of jane's expense$= \frac{300}{0.15} = \$2{,}000$, Percent of jane spend on his car $= 27\%$. Then, jane spend on his car: $27\% \times 2{,}000 = 0.27 \times 2{,}000 = 540$)

2) 200 (Percent of jane spend for foods $= 10\%$. Then, jane spend for foods: $10\% \times 2{,}000 = 0.1 \times 2{,}000 = 200$)

3) 500 (Percent of jane spend on his rent $= 10\%$. Then, jane spend on his rent: $25\% \times 2{,}000 = 0.25 \times 2{,}000 = 500$)

4) $\frac{21}{50}$ (jane's expenses for his bills and Car out of his total expenses $= \frac{Percent\ of\ jane\ spend\ on\ his\ car + Percent\ of\ jane\ spend\ on\ his\ bill}{100} = \frac{15+27}{100} = \frac{42}{100} = \frac{21}{50}$)

Probability Problems

1) $\frac{7}{15}$ (Even numbers between 1 and 15 $\rightarrow \{2, 4, 6, 8, 10, 12, 14\}$.

$Probability = \frac{number\ of\ desired\ outcomes}{number\ of\ total\ outcomes} = \frac{Number\ of\ even\ numbers\ between\ 1\ and\ 15}{number\ of\ total\ outcomes} = \frac{7}{15}$)

2) $\frac{4}{9}, \frac{2}{3}$ (The total number of marbles is 9 and 4 of them are red. Then:

$probability\ of\ red\ marbles = \frac{number\ of\ desired\ outcomes}{number\ of\ total\ outcomes} = \frac{4}{9}$

White marble does not come out means that the marble is yellow or red.

Then: $probability\ that\ white\ marble\ will\ not\ come\ out = \frac{4+2}{9} = \frac{6}{9} = \frac{2}{3}$)

Permutations and Combinations

1) 15 (Since the order doesn't matter, we need to use a combination formula where n is 6 and r is 4. Then: $\frac{n!}{r!\,(n-r)!} = \frac{6!}{4!\,(6-4)!} = \frac{6!}{4!\,(2)!} = \frac{6\times5\times4!}{4!\,(2)!} = \frac{6\times5}{2\times1} = \frac{30}{2} = 15$)

2) 210 (Since the order doesn't matter, we need to use a combination formula where n is 10 and r is 6. Then: $\frac{n!}{r!\,(n-r)!} = \frac{10!}{6!\,(10-6)!} = \frac{10!}{6!\,4!} = \frac{10\times9\times8\times7\times6!}{6!\,4!} = \frac{10\times9\times8\times7}{4\times3\times2\times1} = \frac{5{,}040}{24} = 210$)

3) 2,184 (Since the order matters, we need to use permutation formula where n is 7 and k is 2. Then: $\frac{n!}{(n-k)!} = \frac{14!}{(14-3)!} = \frac{14!}{11!} = \frac{14\times13\times12\times11!}{11!} = 14 \times 13 \times 12 =$ 2,184)

4) 40 (Since the order doesn't matter, we need to use a combination formula where n is 6 and r is 4. Then: $\frac{n!}{r!\,(n-r)!} = \frac{16!}{13!\,(16-13)!} = \frac{16\times15\times13!}{13!\,(3)!} = \frac{16\times15}{3\times2\times1} = \frac{240}{6} = 40$)

CHAPTER

13 Functions Operations

Math topics that you'll learn in this chapter:

☑ Function Notation and Evaluation

☑ Adding and Subtracting Functions

☑ Multiplying and Dividing Functions

☑ Composition of Functions

Function Notation and Evaluation

- Functions are mathematical operations that assign unique outputs to given inputs.
- Function notation is the way a function is written. It is meant to be a precise way of giving information about the function without a rather lengthy written explanation.
- The most popular function notation is $f(x)$ which is read "f of x". Any letter can name a function. for example: $g(x)$, $h(x)$, etc.
- To evaluate a function, plug in the input (the given value or expression) for the function's variable (place holder, x).

Examples:

Example 1. Evaluate: $f(x) = x + 6$, find $f(2)$

Solution: Substitute x with 2:
Then: $f(x) = x + 6 \rightarrow f(2) = 2 + 6 \rightarrow f(2) = 8$

Example 2. Evaluate: $w(x) = 3x - 1$, find $w(4)$.

Solution: Substitute x with 4:
Then: $w(x) = 3x - 1 \rightarrow w(4) = 3(4) - 1 = 12 - 1 = 11$

Practices:

Evaluate each function.

1) $g(n) = 4n - 3$, find $g(4)$ ____

2) $h(x) = 10x + 6$, find $h(2)$ ____

3) $k(n) = 15 - 3n$, find $k(3)$ ____

4) $g(x) = -7x + 5$, find $g(-9)$ ____

5) $k(n) = 12n + 2$, find $k(-5)$ ____

6) $w(n) = -6n - 6$, find $w(7)$ ____

Adding and Subtracting Functions

- Just like we can add and subtract numbers and expressions, we can add or subtract functions and simplify or evaluate them. The result is a new function.
- For two functions $f(x)$ and $g(x)$, we can create two new functions:

$$(f+g)(x)=f(x)+g(x) \text{ and } (f-g)(x)=f(x)-g(x)$$

Examples:

Example 1. $g(x)=2x-2$, $f(x)=x+1$, Find: $(g+f)(x)$

Solution: $(g+f)(x)=g(x)+f(x)$
Then: $(g+f)(x)=(2x-2)+(x+1)=2x-2+x+1=3x-1$

Example 2. $f(x)=4x-3$, $g(x)=2x-4$, Find: $(f-g)(x)$

Solution: $(f-g)(x)=f(x)-g(x)$
Then: $(f-g)(x)=(4x-3)-(2x-4)=4x-3-2x+4=2x+1$

Practices:

Perform the indicated operation.

1) $f(x)=6x-4$
   $g(x)=x+2$
   Find $(f-g)(x)$

   ______

2) $g(x)=5x+5$
   $f(x)=3x-1$
   Find $(g-f)(x)$

   ______

3) $h(t)=8t+10$
   $g(t)=4t+6$
   Find $(h+g)(t)$

   ______

4) $g(a)=6a+7$
   $f(a)=-3a^2-2$
   Find $(g+f)(4)$

   ______

5) $g(x)=12x-4$
   $h(x)=-2x^2+1$
   Find $(g-h)(-2)$

   ______

6) $h(x)=-x-9$
   $g(x)=-x^2-4$
   Find $(h-g)(-6)$

   ______

bit.ly/3hdeFVO
Find more at

Multiplying and Dividing Functions

- Just like we can multiply and divide numbers and expressions, we can multiply and divide two functions and simplify or evaluate them.
- For two functions $f(x)$ and $g(x)$, we can create two new functions:

$$(f.g)(x) = f(x).g(x) \text{ and } \left(\frac{f}{g}\right)(x) = \frac{f(x)}{g(x)}$$

Examples:

Example 1. $g(x) = x + 3$, $f(x) = x + 4$, Find: $(g.f)(x)$

Solution: $(g.f)(x) = g(x).f(x) = (x + 3)(x + 4) = x^2 + 4x + 3x + 12 = x^2 + 7x + 12$

Example 2. $f(x) = x + 6$, $h(x) = x - 9$, Find: $\left(\frac{f}{h}\right)(x)$

Solution: $\left(\frac{f}{h}\right)(x) = \frac{f(x)}{h(x)} = \frac{x+6}{x-9}$

Practices:

Perform the indicated operation.

1) $g(x) = 2x - 1$
   $f(x) = x + 2$
   Find $(g.f)(x)$
   ______

2) $f(x) = x - 3$
   $h(x) = 5x$
   Find $(f.h)(x)$
   ______

3) $g(a) = a + 4$
   $h(a) = 3a - 5$
   Find $(g.h)(2)$
   ______

4) $f(x) = 5x - 4$
   $h(x) = x + 1$
   Find $\left(\frac{f}{h}\right)(-4)$
   ______

5) $f(x) = -6x - 3$
   $g(x) = 4 - 2x$
   Find $\left(\frac{f}{g}\right)(-5)$
   ______

6) $g(a) = -a + 7$
   $f(a) = 2a^2 - 6$
   Find $\left(\frac{g}{f}\right)(6)$
   ______

Composition of Functions

- "Composition of functions" simply means combining two or more functions in a way where the output from one function becomes the input for the next function.
- The notation used for composition is: $(fog)(x) = f(g(x))$ and is read "f composed with g of x" or "f of g of x".

Examples:

Example 1. Using $f(x) = 2x + 3$ and $g(x) = 5x$, find: $(fog)(x)$

Solution: $(fog)(x) = f(g(x))$. Then: $(fog)(x) = f(g(x)) = f(5x)$

Now find $f(5x)$ by substituting x with $5x$ in $f(x)$ function.

Then: $f(x) = 2x + 3$; $(x \rightarrow 5x) \rightarrow f(5x) = 2(5x) + 3 = 10x + 3$

Example 2. Using $f(x) = 2x^2 - 5$ and $g(x) = x + 3$, find: $f(g(3))$

Solution: First, find $g(3)$: $g(x) = x + 3 \rightarrow g(3) = 3 + 3 = 6$

Then: $f(g(3)) = f(6)$. Now, find $f(6)$ by substituting x with 6 in $f(x)$ function.

$f(g(3)) = f(6) = 2(6)^2 - 5 = 2(36) - 5 = 67$

Practices:

Using $f(x) = 2x + 6$ and $g(x) = x - 4$, find:

1) $g(f(3)) =$_____
2) $g(f(6)) =$_____
3) $f(g(-7)) =$_____
4) $f(f(10)) =$_____
5) $g(f(-4)) =$_____
6) $g(f(7)) =$_____
7) $g(f(8)) =$ ____
8) $g(f(5)) =$ ____

bit.ly/2WHBkAg

Find more at

Chapter 13: Answers

Function Notation and Evaluation

1) 13 (Substitute n with 4: Then: $g(n) = 4n - 3 \rightarrow g(4) = 4(4) - 3 \rightarrow g(4) = 16 - 3 \rightarrow g(4) = 13$)
2) 26 (Substitute x with 2: Then:$h(x) = 10x + 6 \rightarrow h(2) = 10(2) + 6 \rightarrow h(2) = 20 + 6 \rightarrow h(2) = 26$)
3) 6 (Substitute n with 3: Then: $k(n) = 15 - 3n \rightarrow k(3) = 15 - 3(3) \rightarrow k(3) = 15 - 9 \rightarrow k(3) = 6$)
4) 68 (Substitute x with -9: Then: $g(x) = -7x + 5 \rightarrow g(-9) = -7(-9) + 5 \rightarrow g(-9) = 63 + 5 \rightarrow g(-9) = 68$)
5) -58 (Substitute n with -5: Then: $k(n) = 12n + 2 \rightarrow k(-5) = 12(-5) + 2 \rightarrow k(-5) = -60 + 2 \rightarrow k(-5) = -58$)
6) -48 (Substitute n with 7: Then: $w(n) = -6n - 6 \rightarrow w(7) = -6(7) - 6 \rightarrow w(7) = -42 - 6 \rightarrow w(7) = -48$)

Adding and Subtracting Functions

1) $5x - 6$ ($(f - g)(x) = f(x) - g(x)$. Then: $(f - g)(x) = (6x - 4) - (x + 2) = 6x - 4 - x - 2 = 5x - 6$)
2) $2x + 6$ ($(g - f)(x) = g(x) - f(x)$. Then: $(g - f)(x) = (5x + 5) - (3x - 1) = 5x + 5 - 3x + 1 = 2x + 6$)
3) $12t + 16$ ($(h + g)(t) = h(t) + g(t)$. Then: $(h + g)(t) = (8t + 10) + (4t + 6) = 8t + 10 + 4t + 6 = 12t + 16$)
4) -19 ($(g + f)(a) = g(a) + f(a)$. Then: $(g + f)(a) = (6a + 7) + (-3a^2 - 2) = 6a + 7 - 3a^2 - 2 = -3a^2 + 6a + 5$. Substitute a with 4: $(g + f)(4) = -3(4)^2 + 6(4) + 5 = -48 + 24 + 5 = -19$)

5) -21 ($(g-h)(x) = g(x) - h(x)$. Then: $(g-h)(x) =$ $(12x-4) - (-2x^2+1) = 12x - 4 + 2x^2 - 1 = 2x^2 + 12x - 5$. Substitute x with -2: $(g-h)(-2) = 2(-2)^2 + 12(-2) - 5 = 8 - 24 - 5 = -21$)

6) 37 ($(h-g)(x) = h(x) - g(x)$. Then: $(h-g)(x) =$ $(-x-9) - (-x^2-4) = -x - 9 + x^2 + 4 = x^2 - x - 5$. Substitute x with -6: $(h-g)(-6) = (-6)^2 - (-6) - 5 = 36 + 6 - 5 = 37$)

Multiplying and Dividing Functions

1) $2x^2 + 3x - 2$ ($(g.f)(x) = g(x).f(x) = (2x-1)(x+2) =$ $2x^2 + 4x - x - 2 = 2x^2 + 3x - 2$)

2) $5x^2 - 15x$ ($(f.h)(x) = f(x).h(x) = (x-3)(5x) = 5x^2 - 15x$)

3) 6 ($(g.h)(a) = g(a).h(a) = (a+4)(3a-5) = 3a^2 - 5a + 12a - 20$ $g(a).h(a) = 3a^2 + 7a - 20$. Substitute a with 2: $(g.h)(2) =$ $3(2)^2 + 7(2) - 20 = 12 + 14 - 20 = 6$)

4) 8 ($\left(\frac{f}{h}\right)(x) = \frac{f(x)}{h(x)} = \frac{5x-4}{x+1}$. Substitute x with -4: $\left(\frac{f}{h}\right)(-4) = \frac{5x-4}{x+1} = \frac{5(-4)-4}{(-4)+1} = \frac{-24}{-3} = 8$)

5) $-\frac{27}{6}$ ($\left(\frac{f}{g}\right)(x) = \frac{f(x)}{g(x)} = \frac{-6x-3}{4-2x}$. Substitute x with -5: $\left(\frac{f}{g}\right)(-5) = \frac{-6x-3}{4-2x} = \frac{-6(-5)-3}{4-2(-5)} = \frac{27}{-6} = -\frac{27}{6}$)

6) $\frac{1}{66}$ ($\left(\frac{g}{f}\right)(a) = \frac{g(a)}{f(a)} = \frac{-a+7}{2a^2-6}$. Substitute a with 6: $\left(\frac{g}{f}\right)(6) = \frac{-a+7}{2a^2-6} = \frac{-(6)+7}{2(6)^2-6} = \frac{1}{66}$)

Composition of Functions

1) 8 (First, find $f(3)$: $f(x) = 2x + 6 \rightarrow f(3) = 2(3) + 6 = 6 + 6 = 12$. Then: $g(f(3)) = g(12)$. Now, find $g(12)$ by substituting x with 12 in $g(x)$ function. $g(f(3)) = g(x) = x - 4 \rightarrow g(12) = 12 - 4 = 8$)

2) 14 (First, find $f(6)$: $f(x) = 2x + 6 \rightarrow f(6) = 2(6) + 6 = 12 + 6 = 18$. Then: $g(f(6)) = g(18)$. Now, find $g(18)$ by substituting x with 18 in $g(x)$ function. $g(f(6)) = g(x) = x - 4 \rightarrow g(18) = 18 - 4 = 14$)

3) -16 (First, find $g(-7)$: $g(x) = x - 4 \rightarrow g(-7) = -7 - 4 = -11$. Then: $f(g(-7)) = f(-11)$. Now, find $f(-11)$ by substituting x with -11 in $f(x)$ function. $f(g(-7)) = f(x) = 2x + 6 \rightarrow f(-11) = 2(-11) + 6 = -22 + 6 = -16$)

4) 58 (First, find $f(10)$: $f(x) = 2x + 6 \rightarrow f(10) = 2(10) + 6 = 20 + 6 = 26$. Then: $f(f(10)) = f(26)$. Now, find $f(26)$ by substituting x with 26 in $f(x)$ function. $f(f(10)) = f(x) = 2x + 6 \rightarrow f(26) = 2(26) + 6 = 58$)

5) -6 (First, find $f(-4)$: $f(x) = 2x + 6 \rightarrow f(-4) = 2(-4) + 6 = -8 + 6 = -2$. Then: $g(f(-4)) = g(-2)$. Now, find $g(-2)$ by substituting x with -2 in $g(x)$ function. $g(f(-4)) = g(x) = x - 4 \rightarrow g(-2) = -2 - 4 = -6$)

6) 16 (First, find $f(7)$: $f(x) = 2x + 6 \rightarrow f(7) = 2(7) + 6 = 14 + 6 = 20$. Then: $g(f(7)) = g(20)$. Now, find $g(20)$ by substituting x with 20 in $g(x)$ function. $g(f(7)) = g(x) = x - 4 \rightarrow g(20) = 20 - 4 = 16$)

7) 18 (First, find $f(8)$: $f(x) = 2x + 6 \rightarrow f(8) = 2(8) + 6 = 16 + 6 = 22$. Then: $g(f(8)) = g(22)$. Now, find $g(22)$ by substituting x with 22 in $g(x)$ function. $g(f(8)) = g(x) = x - 4 \rightarrow g(22) = 22 - 4 = 18$)

8) 12 (First, find $f(5)$: $f(x) = 2x + 6 \rightarrow f(5) = 2(5) + 6 = 10 + 6 = 16$. Then: $g(f(5)) = g(16)$. Now, find $g(16)$ by substituting x with 16 in $g(x)$ function. $g(f(5)) = g(x) = x - 4 \rightarrow g(16) = 16 - 4 = 12$)

Time to test

Time to refine your skill with a practice examination

In this section, there are two complete Praxis Core Math Tests. Take these tests to simulate the test day experience. After you've finished, score your test using the answer keys.

Before You Start

- You'll need a pencil, a calculator and a timer to take the test.
- For each question, there are five possible answers. Choose which one is best.
- It's okay to guess. There is no penalty for wrong answers.
- After you've finished the test, review the answer key to see where you went wrong.

Good luck!

Praxis Core Practice Test 1

2023

56 questions

Total time for this section: 90 minutes

<u>You may use a calculator on this practice test.</u>

(On a real Praxis test, there is an onscreen calculator to use.)

Formula Sheet

Perimeter / Circumference

Rectangle

$Perimeter = 2(length) + 2(width)$

Circle

$Circumference = 2\pi(radius)$

Area

Circle

$Area = \pi(radius)^2$

Triangle

$Area = \frac{1}{2}(base)(height)$

Parallelogram

$Area = (base)(height)$

Trapezoid

$Area = \frac{1}{2}(base_1 + base_2)(height)$

Volume

Prism/Cylinder

$Volume = (area\ of\ the\ base)(height)$

Pyramid/Cone

$Volume = \frac{1}{3}(area\ of\ the\ base)(height)$

Sphere

$Volume = \frac{4}{3}\pi(radius)^3$

Length

1 foot = 12 inches

1 yard = 3 feet

1 mile = 5,280 feet

1 meter = 1,000 millimeters

1 meter = 100 centimeters

1 kilometer = 1,000 meters

1 mile ≈ 1.6 kilometers

1 inch = 2.54 centimeters

1 foot ≈ 0.3 meter

Capacity / Volume

1 cup = 8 fluid ounces

1 pint = 2 cups

1 quart = 2 pints

1 gallon = 4 quarts

1 gallon = 231 cubic inches

1 liter = 1,000 milliliters

1 liter ≈ 0.264 gallon

Weight

1 pound = 16 ounces

1 ton = 2,000 pounds

1 gram = 1,000 milligrams

1 kilogram = 1,000 grams

1 kilogram ≈ 2.2 pounds

1 ounce ≈ 28.3 grams

1) The capacity of a red box is 20% bigger than the capacity of a blue box. If the red box can hold 30 equal sized books, how many of the same books can the blue box hold?

A. 9
B. 15
C. 21
D. 25
E. 30

2) Kim spent $35 for pants. This was $10 less than triple what she spent for a shirt. How much was the shirt?

A. $11
B. $13
C. $15
D. $17
E. $21

3) What is the greatest integer less than $-\frac{32}{5}$?

A. 0
B. -2
C. -4
D. -6
E. -7

4) The measure of the angles of a triangle are in the ratio $1:3:5$. What is the measure of the largest angle?

A. 20°
B. 45°
C. 85°
D. 100°
E. 180°

5) In the figure below, line A is parallel to line B. what is the value of x?

A. 28
B. 46
C. 50
D. 55
E. 65

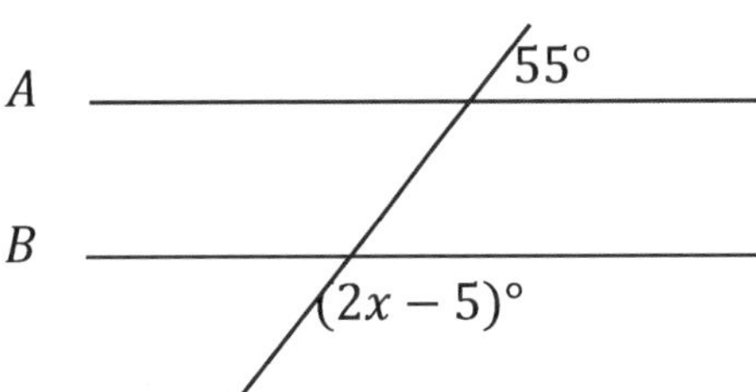

6) In the infinitely repeating decimal below, 1 is the first digit in the repeating pattern. What is the $68th$ digit? $\frac{1}{7} = 0.\overline{142857}$

A. 1
B. 2
C. 4
D. 5
E. 7

7) Which of the following answers represents the compound inequality?

$$-4 \leq 4x - 8 < 16$$

A. $-2 \leq x \leq 8$
B. $-2 < x \leq 8$
C. $1 < x \leq 6$
D. $1 \leq x < 6$
E. $2 \leq x \leq 6$

8) In the following figure, $ABCD$ is a rectangle. If $a = \sqrt{3}$, and $b = 2a$, find the area of the shaded region. (the shaded region is a trapezoid)

A. $2\sqrt{3}$
B. $3\sqrt{3}$
C. $4\sqrt{3}$
D. $6\sqrt{3}$
E. $8\sqrt{3}$

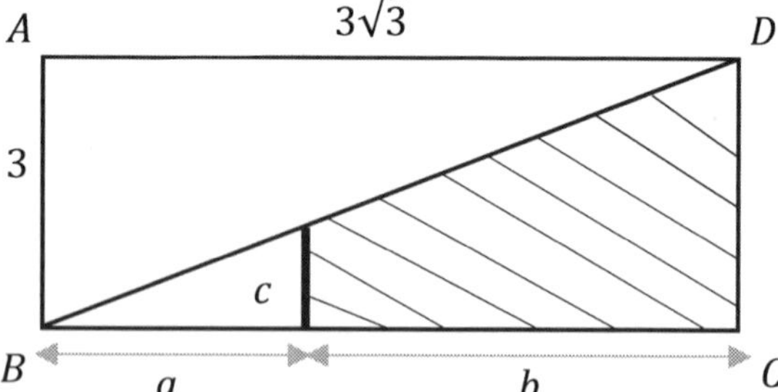

9) In the figure below, what is the value of x?

Write your answer in the box below.

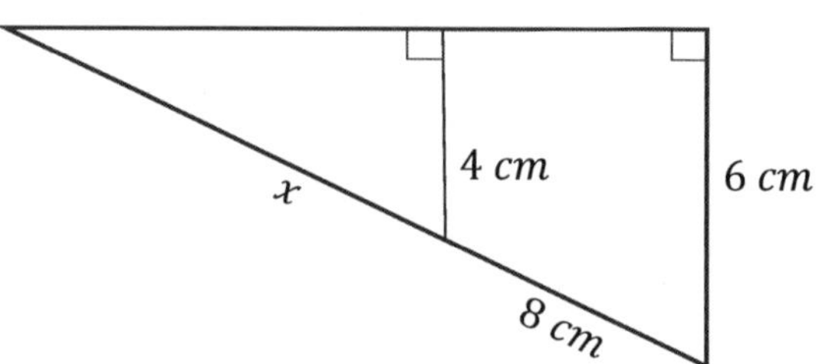

10) Which graph shows a non-proportional linear relationship between x and y?

A.

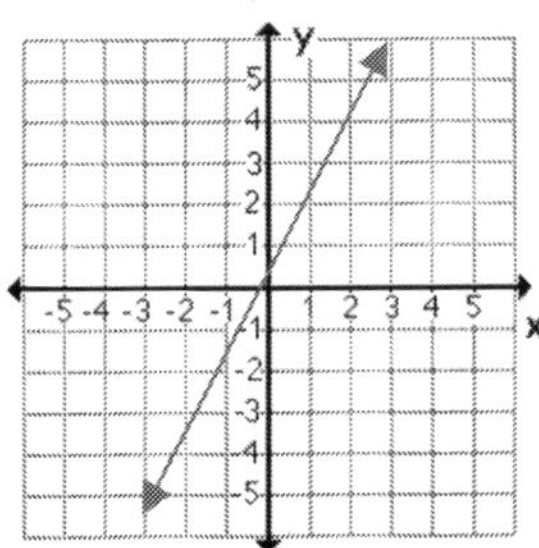

B.

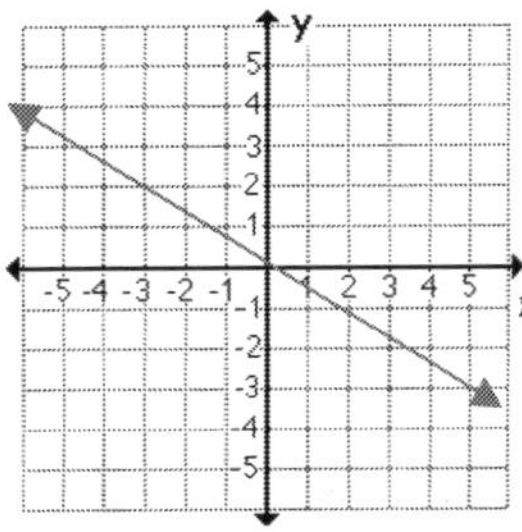

C.

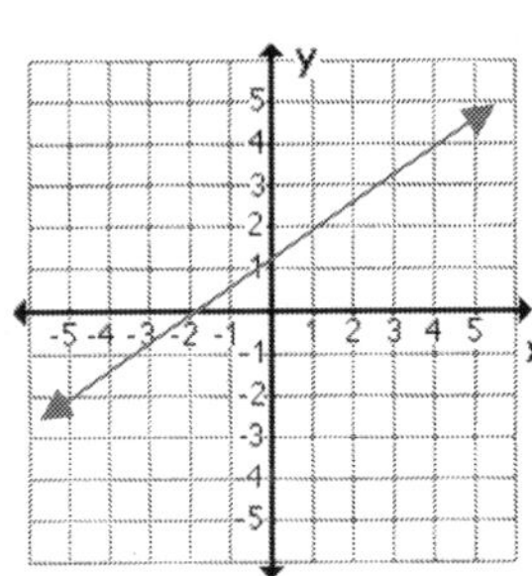

D.

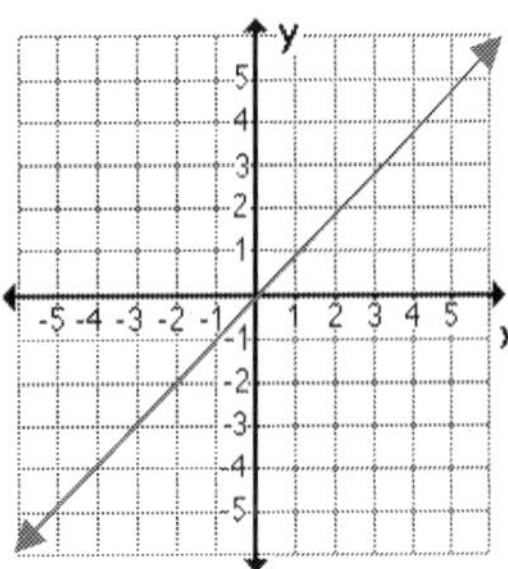

E.

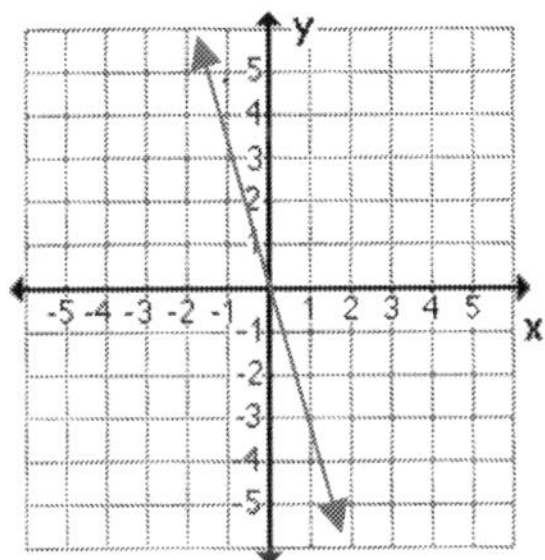

11) Anna opened an account with a deposit of $3,000. This account earns 5% simple interest annually. How many years will it take her to earn $600 on her $3,000 deposit?

A. 2
B. 4
C. 5
D. 6
E. 8

12) Tom picked $2\frac{2}{5}$ baskets of apples, and Sam picked $1\frac{3}{4}$ baskets of apples. How many baskets total did they pick?

A. $1\frac{2}{3}$
B. $2\frac{1}{12}$
C. $3\frac{20}{23}$
D. $4\frac{3}{20}$
E. $5\frac{1}{12}$

13) A list of consecutive integers begins with k and ends with n. If $n-k=46$, how many integers are in the list?

A. 23
B. 38
C. 46
D. 47
E. 58

14) A piece of paper that is $2\frac{3}{5}$ feet long is cut into 2 pieces of different lengths. The shorter piece has a length of x feet. Which inequality expresses all possible values of x?

A. $x < 2\frac{1}{10}$
B. $x > 2$
C. $x < 2\frac{3}{5}$
D. $x > 1\frac{3}{10}$
E. $x < 1\frac{3}{10}$

15) In an academy, course grades range from 0 to 100. Anna took 5 courses and her mean course grade was 80. William took 8 courses. If both students have the same sum of course grades, what was William's mean?

A. 50
B. 65
C. 70
D. 80
E. 85

16) The sum of the numbers x, y, and z is 69. The ratio of x to y is $1:3$ and the ratio of y to z is $2:5$. What is the value of y?

A. 9.6
B. 15
C. 18
D. 21.6
E. 33

17) Which number line below shows the solution to the inequality $-1 < \frac{x}{3} < 2$?

A. 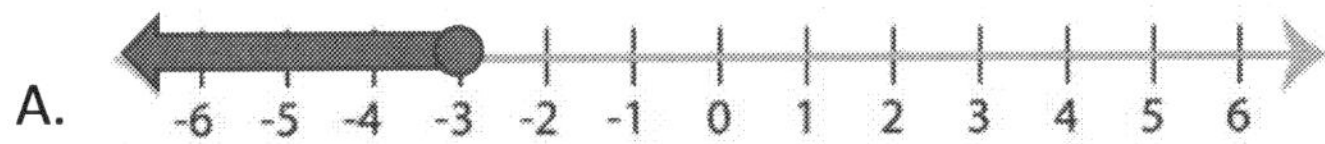

B. 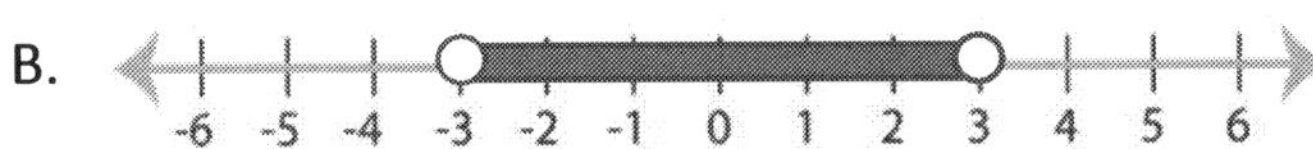

C. 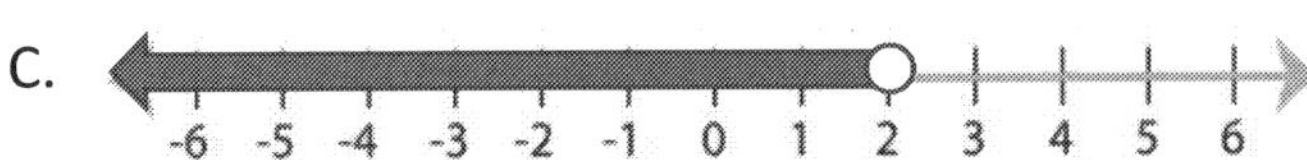

D. 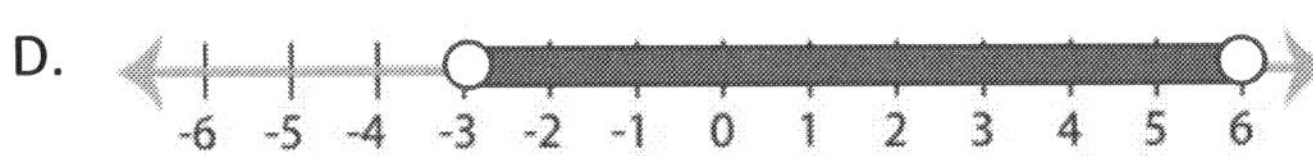

E. 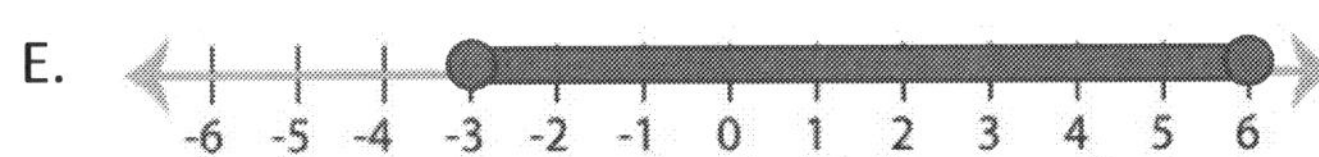

18) The set S consists of all odd numbers greater than 5 and less than 30. What is the mean of the numbers in S.

A. 11
B. 13
C. 17
D. 18
E. 23

19) If $\sqrt{2y} = \sqrt{5x}$ then $x = \cdots$

A. $\frac{1}{6}y$
B. $\frac{1}{5}y$
C. $\frac{2}{5}y$
D. $\frac{5}{2}y$
E. $10y$

20) A line connects the midpoint of AB (point E), with point C in the square $ABCD$. Calculate the area of the acquired trapezoid shape if the square has a side of $4\ m$.

A. $4\ cm^2$
B. $12\ cm^2$
C. $15\ cm^2$
D. $18\ cm^2$
E. $24\ cm^2$

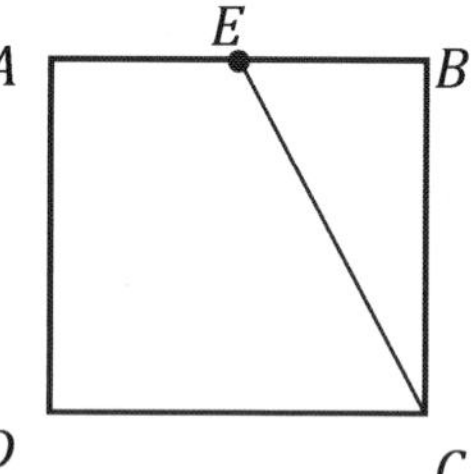

21) In a group of 45 student, 60% can't swim. How many students can swim?

A. 13
B. 18
C. 22
D. 23
E. 35

22) $\$8{,}400$ are distributed equally among 14 person. How much money will each person get?

A. $\$400$
B. $\$450$
C. $\$584$
D. $\$600$
E. $\$800$

23) A box contains 6 green sticks, 4 blue sticks, and 2 yellow sticks. Emma picks one without looking. What is the probability that the stick will be green?

A. $\frac{1}{2}$
B. $\frac{1}{3}$
C. $\frac{1}{4}$
D. $\frac{2}{5}$
E. $\frac{3}{2}$

24) The figure below, a square is inscribed in a circle. Calculate the shaded area in the figure below. Knowing that the radius of the circle is $6\ cm$. ($\pi = 3.14$)

A. $73.65\ cm^2$
B. $69.90\ cm^2$
C. $72.69\ cm^2$
D. $88.04\ cm^2$
E. $113.4\ cm^2$

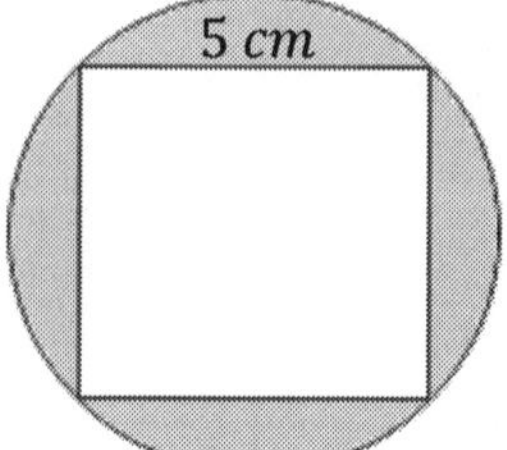

25) The price of a Chocolate was raised from $\$5.40$ to $\$5.67$. What was the percent increase in the price?

A. 4%
B. 5%
C. 6%
D. 8%
E. 10%

26) In a box of blue and black marbles, the ratio of blue marbles to black marbles is $4:3$. If the box contains 150 black marbles, how many blue marbles are there?

A. 100
B. 150
C. 200
D. 300
E. 600

27) $\frac{5}{8}$ of a number is 90. Find the number.

A. 144
B. 270
C. 450
D. 720
E. 800

28) A juice mixture contains $\frac{5}{14}$ jar of cherry juice and $\frac{5}{70}$ jar of apple juice. How many jars of cherry juice per jar of apple juice does the mixture contain?

A. 70
B. 14
C. 10
D. 7
E. 5

29) In the figure below, $LMNO$ and $JMPQ$ are squares. Point O is the center of the circle, and points L and N are on the circle. If the area of the square is 16 square centimeters, what is the area, in square centimeters, of the shaded part?

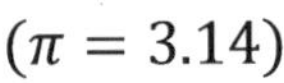
$(\pi = 3.14)$

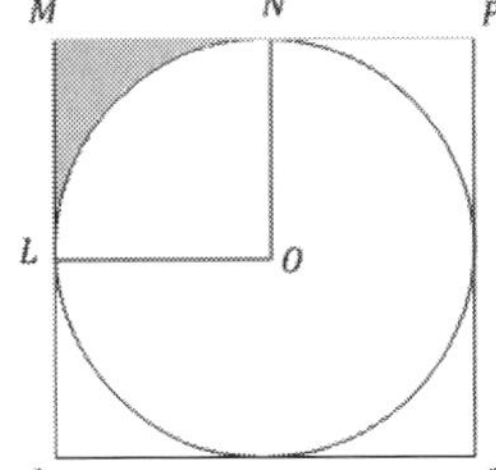

Write your answer in the box below.

30) The set of possible values of n is $\{5, 3, 7\}$. What is the set of possible values of m if $2m = n + 5$?

A. $\{2, 4, 7\}$
B. $\{3, 2, 5\}$
C. $\{4, 5, 8\}$
D. $\{5, 4, 6\}$
E. $\{6, 5, 8\}$

31) If $x = 25$, then which of the following equations are correct?

A. $x + 10 = 40$
B. $4x = 100$
C. $3x = 70$
D. $\frac{x}{2} = 12$
E. $\frac{x}{3} = 8$

32) Jack scored a mean of 80 per test in his first 4 tests. In his 5th test, he scored 90. What was Jack's mean score for the 5 tests?

A. 70
B. 75
C. 80
D. 82
E. 93

33) The volume of a cube is less than $64\ m^3$. Which of the following can be the cube's side?

A. $2\ m$
B. $4\ m$
C. $8\ m$
D. $10\ m$
E. $11\ m$

34) What is the area of an isosceles right triangle that has one leg that measures $6\ cm$?

Write your answer in the box below.

35) If $0.00104 = \frac{104}{x}$, what is the value of x?

A. 1,000
B. 10,000
C. 100,000
D. 1,000,000
E. 10,000,000

36) A bag is filled with numbered cards from 1 to 15 and picked on at random. What is the probability that the card picked is number 8?

A. $\frac{8}{15}$
B. $\frac{7}{15}$
C. $\frac{5}{15}$
D. $\frac{2}{15}$
E. $\frac{1}{15}$

37) In the xy-plane, the point $(4, 3)$ and $(3, 2)$ are on line A. Which of the following points could also be on line A?

A. $(5, 7)$
B. $(3, 4)$
C. $(-1, 2)$
D. $(-1, -2)$
E. $(-7, -9)$

38) If $f(x)=2x^3+ 5x^2+ 2x$ and $g(x)= -2$, what is the value of $f(g(x))$?

A. 36
B. 32
C. 24
D. 4
E. 0

39) What is the value of x in the figure below?

Write your answer in the box below.

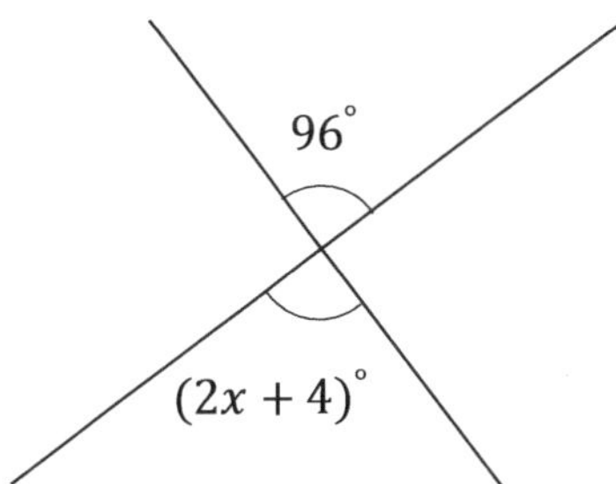

40) How many different two-digit numbers can be formed from the digits $6, 7$, and 5, if the numbers must be even and no digit can be repeated?

A. 1
B. 2
C. 3
D. 4
E. 5

41) A rectangular concrete driveway is 25 feet long, 6 feet wide, and 24 inches thick. What is the volume of the concrete?

A. $300\ ft^3$
B. $600\ ft^3$
C. $660\ ft^3$
D. $963\ ft^3$
E. $1{,}800\ ft^3$

42) If $\frac{2y}{x} - \frac{y}{3x} = \frac{(\ldots)}{3x}$ and $x \neq 0$, what expression is represented by $(\ldots)$?

A. $2y + 4$
B. $3y - 6$
C. $5y$
D. $6y$
E. $8y$

43) $200(3 + 0.01)^2 - 200 =$

A. 201.55
B. 361.08
C. 702.88
D. 1,612.02
E. 1,812.02

44) If 360 kg of vegetables is packed in 90 boxes, how much vegetables will each box contain?

A. 2.5 kg
B. 3 kg
C. 4 kg
D. 6.5 kg
E. 7 kg

45) Each number in a sequence is 4 more than twice the number that comes just before it. If 84 is a number in the sequence, what number comes just before it?

A. 26
B. 35
C. 40
D. 52
E. 88

46) $[6 \times (-24) + 8] - (-4) + [4 \times 5] \div 2 = ?$

A. 148
B. 132
C. -122
D. -136
E. -144

47) Solve for x: $2 + \frac{3x}{x-5} = \frac{3}{5-x}$?

A. $\frac{4}{5}$
B. $\frac{6}{5}$
C. $\frac{7}{5}$
D. $\frac{8}{5}$
E. $\frac{9}{5}$

48) A rectangle has 14 cm wide and 5 cm length. What is the perimeter of this rectangle?

Write your answer in the box below.

49) What is the value of the following expression? $3\frac{1}{4} + 2\frac{4}{16} + 1\frac{3}{8} + 5\frac{1}{2}$

A. $3\frac{10}{14}$
B. $4\frac{1}{2}$
C. $12\frac{4}{16}$
D. $12\frac{3}{8}$
E. $12\frac{4}{8}$

50) A certain insect has a mass of 85 milligrams. What is the insect's mass in grams?

A. 0.085
B. 0.08
C. 0.85
D. 8.5
E. 85

51) Removing which of the following numbers will change the average of the numbers to 6?

$$1, 4, 5, 8, 11, 12$$

A. 1
B. 4
C. 5
D. 8
E. 11

52) If $m = 6$ and $n = -3$, what is the value of $\frac{5-9(3+n)}{3m-5(2-n)} =$?

A. $\frac{2}{7}$
B. $\frac{3}{7}$
C. $-\frac{4}{7}$
D. $\frac{5}{7}$
E. $-\frac{5}{7}$

53) Clara has 28 cookies. She is inviting 7 friends to a party. How many cookies will each friend get?

A. 2
B. 4
C. 7
D. 8
E. 21

54) How long will it take to receive $360 in investment of $240 at the rate of 10% simple interest?

A. 9 years
B. 15 years
C. 18 years
D. 21 years
E. 24 years

55) How many hours are there in 1,800 minutes?

A. 20 hours
B. 25 hours
C. 30 hours
D. 33 hours
E. 60 hours

56) What is the value of x in the following equation? $|34-78|-x+|-12+20|=36$

A. 0
B. 16
C. 23
D. 33
E. 72

End of Praxis Mathematics Practice Test 1

Praxis Core Practice Test 2

2023

56 questions

Total time for this section: 90 minutes

<u>**You may use a calculator on this practice test.**</u>

(On a real Praxis test, there is an onscreen calculator to use.)

Formula Sheet

Perimeter / Circumference

Rectangle

$Perimeter = 2(length) + 2(width)$

Circle

$Circumference = 2\pi(radius)$

Area

Circle

$Area = \pi(radius)^2$

Triangle

$Area = \frac{1}{2}(base)(height)$

Parallelogram

$Area = (base)(height)$

Trapezoid

$Area = \frac{1}{2}(base_1 + base_2)(height)$

Volume

Prism/Cylinder

$Volume = (area\ of\ the\ base)(height)$

Pyramid/Cone

$Volume = \frac{1}{3}(area\ of\ the\ base)(height)$

Sphere

$Volume = \frac{4}{3}\pi(radius)^3$

Length

1 foot = 12 inches

1 yard = 3 feet

1 mile = 5,280 feet

1 meter = 1,000 millimeters

1 meter = 100 centimeters

1 kilometer = 1,000 meters

1 mile ≈ 1.6 kilometers

1 inch = 2.54 centimeters

1 foot ≈ 0.3 meter

Capacity / Volume

1 cup = 8 fluid ounces

1 pint = 2 cups

1 quart = 2 pints

1 gallon = 4 quarts

1 gallon = 231 cubic inches

1 liter = 1,000 milliliters

1 liter ≈ 0.264 gallon

Weight

1 pound = 16 ounces

1 ton = 2,000 pounds

1 gram = 1,000 milligrams

1 kilogram = 1,000 grams

1 kilogram ≈ 2.2 pounds

1 ounce ≈ 28.3 grams

1) A shoe originally priced at $45.00 was on sale for 15% off. Nick received a 20% employee discount applied to the sale price. How much did Nick pay for the shoes?

A. $30.60
B. $34.50
C. $37.30
D. $38.25
E. $42.25

2) In a class, there are 18 boys and 12 girls. What is the ratio of the number of boys to number of girls?

A. 1: 2
B. 1: 3
C. 2: 3
D. 3: 1
E. 3: 2

3) The shaded sector of the circle shown below has an area of 12π square feet. What is the circumference of the circle?

30°

A. $24\pi\ feet$
B. $36\pi\ feet$
C. $81\pi\ feet$
D. $124\pi\ feet$
E. $180\pi\ feet$

4) Which of the following is a factor of 45?

A. 7
B. 9
C. 11
D. 13
E. 14

5) By what percent did the price of a shirt increase if its price was increased from $15.30 to $18.36?

A. 10%
B. 12%
C. 16%
D. 20%
E. 22%

6) The greatest common factor of 32 and x is 8. How many possible values for x are greater than 10 and less than 60?

A. 1
B. 4
C. 6
D. 7
E. 9

7) A box contains 6 strawberry candies, 4 orange candies, and 3 banana candies. If Roberto selects 2 candies at random from this box, without replacement, what is the probability that both candies are not orange?

A. $\frac{1}{28}$
B. $\frac{2}{13}$
C. $\frac{6}{13}$
D. $\frac{1}{3}$
E. $\frac{2}{3}$

8) How many integers are between $\frac{7}{2}$ and $\frac{30}{4}$?

A. 3
B. 4
C. 6
D. 10
E. 12

9) In a certain state, the sales tax rate increased from 8% to 8.5%. What was the increase in the sales tax on a $\$250$ item?

A. $\$0.5$
B. $\$1.00$
C. $\$1.25$
D. $\$1.90$
E. $\$2.30$

10) Triangle ABC is graphed on a coordinate grid with vertices at $A\ (-3,-2)$, $B\ (-1,4)$ and $C\ (7,9)$. Triangle ABC is reflected over x axes to create triangle $A'\ B'\ C'$. Which order pair represents the coordinate of C'?

A. $(-7,-9)$
B. $(-7,9)$
C. $(7,-9)$
D. $(7,9)$
E. $(9,7)$

11) Which of the following is the solution of the following inequality?

$$2x + 4 > 11x - 12.5 - 3.5x$$

A. $x < 3$
B. $x \leq 3$
C. $x > 3$
D. $x \leq 4$
E. $x \geq 4$

12) What is the volume of the following triangular prism?

Write your answer in the box below. (don't write the measurement)

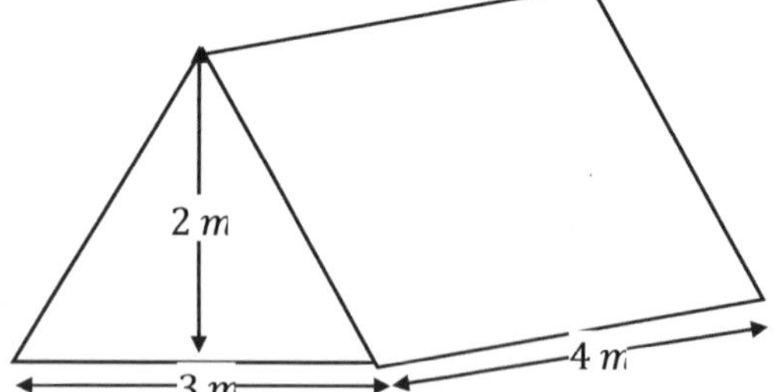

13) Made a list of all possible products of 2 different numbers in the set below. What fraction of the products are odd?

$$\{1, 4, 6, 5, 7\}$$

A. $\frac{2}{5}$
B. $\frac{3}{10}$
C. $\frac{7}{10}$
D. $\frac{4}{15}$
E. $\frac{8}{17}$

14) Which of the following equations has a graph that is a straight line?

A. $y = 3x^2 + 9$
B. $x^2 + y^2 = 1$
C. $4x - 2y = 2x$
D. $7x + 2xy = 6$
E. $x + 2 = y^2$

15) If $5n$ is a positive even number, how many odd numbers are in the range from $5n$ up to and including $5n + 6$?

A. 1
B. 2
C. 3
D. 4
E. 5

16) What is the value of x in the following equation?

$$\frac{2}{3}x + \frac{1}{6} = \frac{1}{3}$$

A. 6
B. $\frac{1}{2}$
C. $\frac{1}{3}$
D. $\frac{1}{4}$
E. $\frac{1}{12}$

17) A bank is offering 2.5% simple interest on a savings account. If you deposit \$16,000, how much interest will you earn in three years?

A. \$610
B. \$1,200
C. \$2,400
D. \$4,800
E. \$6,400

18) If $b = 2$ and $\frac{a}{4} = b$, what is the value of $a^2 + 4b$?

A. 54
B. 66
C. 72
D. 76
E. 81

19) For what value of x is the proportion true? $x: 40 = 20: 32$

A. 16
B. 25
C. 28
D. 32
E. 36

20) When asked a certain question in a poll, 76% of the people polled answered yes. If 66 people did not answer yes to that question, what is the total number of people who were polled?

Write your answer in the box below.

21) Which percentage is closest in value to 0.0099?

A. 0.1%
B. 1%
C. 2%
D. 9%
E. 100%

22) What is the surface area of the cylinder below?

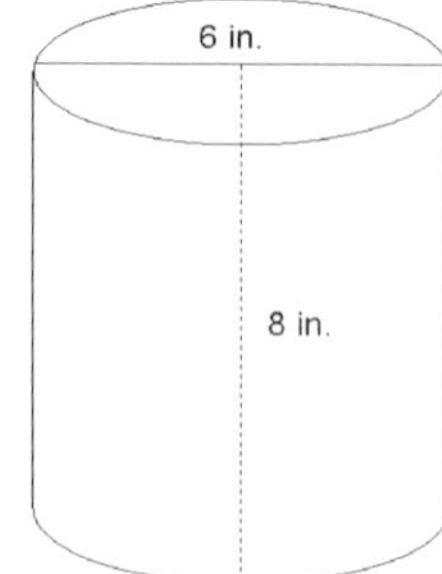

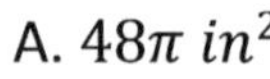

A. $48\pi\ in^2$
B. $57\pi\ in^2$
C. $66\pi\ in^2$
D. $288\pi\ in^2$
E. $400\pi\ in^2$

23) A train travels 1,500 miles from New York to Oklahoma. The train covers the first 280 miles in 4 hours. If the train continues to travel at this rate, how many more hours will it take to reach Oklahoma City? Round your answer to the nearest whole hour.

A. 12
B. 15
C. 17
D. 20
E. 22

24) Which of the following graphs represents the solution of $|9 + 6x| \leq 3$?

A.

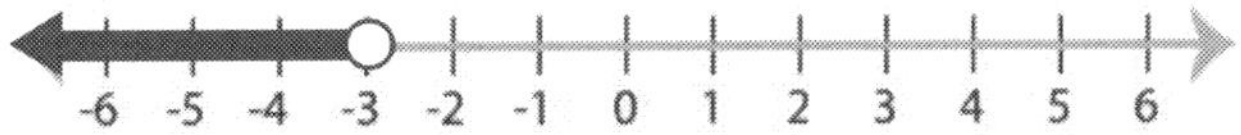

B.

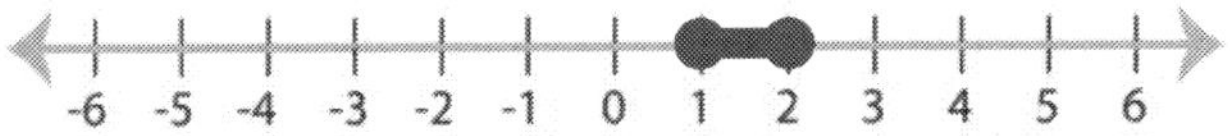

C.

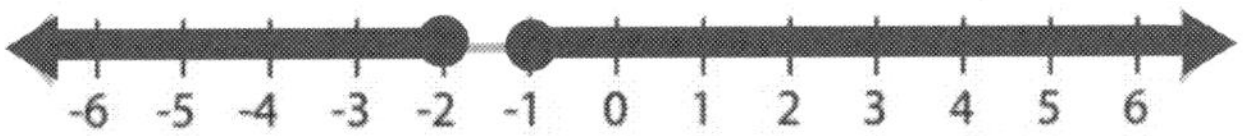

D.

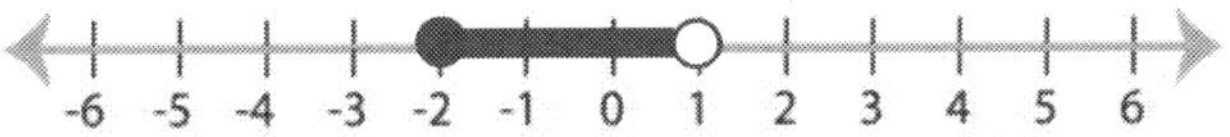

E.

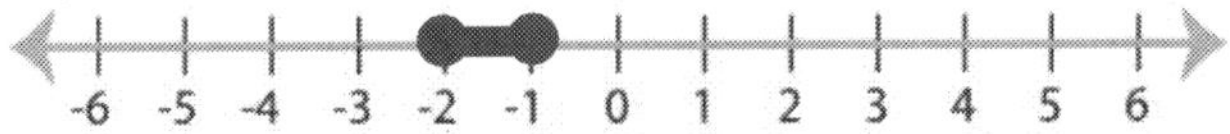

25) In the xy −plane, the line determined by the points $(6, m)$ and $(m, 12)$ passes through the origin. Which of the following could be the value of m?

A. 6
B. 9
C. 12
D. $\sqrt{6}$
E. $6\sqrt{2}$

26) What is the average of the circumference of figure A and the area of figure B? $(\pi = 3)$

Write your answer in the box below.

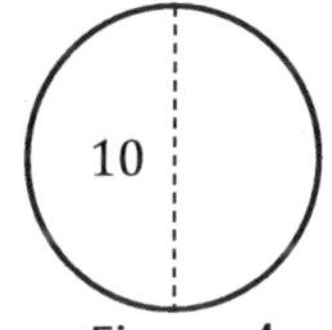

Figure A

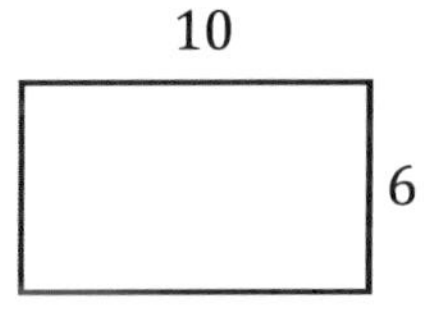

Figure B

27) The sales price of a laptop is $1,912.50, which is 15% off the original price. What is the original price of the laptop?

A. 2,750
B. 2,250
C. 1,625.625
D. 956.25
E. 286.875

28) The diameter of the given semi−circular slice is $18\ cm$. What is the perimeter of the slice? $(\pi = 3.14)$

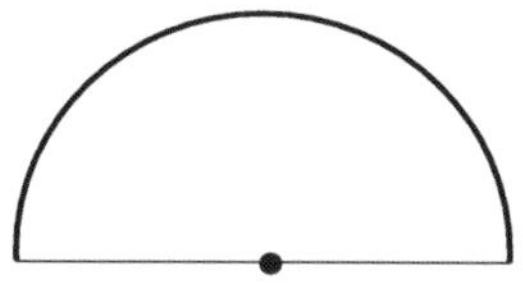

A. $56.52\ cm$
B. $46.26\ cm$
C. $27\ cm$
D. $18\ cm$
E. $9\ cm$

29) In a scale diagram, 0.15 inch represents 150 feet. How many inches represent $2.5\ feet$?

A. $0.001\ in$
B. $0.002\ in$
C. $0.0025\ in$
D. $0.01\ in$
E. $0.012\ in$

30) If $\frac{3}{7}$ of Z is 54, what is $\frac{2}{5}$ of Z?

A. 44.2
B. 46.3
C. 48.4
D. 50.4
E. 60.6

31) A car travels at a speed of 72 miles per hour. How far will it travel in 8 hours?

A. 576
B. 540
C. 480
D. 432
E. 272

32) If Sam spent $\$60$ on sweets and he spent 25% of the selling price for the tip, how much did he spend?

A. $\$66$
B. $\$69$
C. $\$72$
D. $\$75$
E. $\$77$

33) Which of the following numbers has factors that include the smallest factor (other than 1) of 95?

A. 25
B. 28
C. 32
D. 39
E. 45

34) $\frac{4^2+3^2+(-5)^2}{(9+10-11)^2}=?$

A. $\frac{25}{32}$
B. $-\frac{25}{32}$
C. 56
D. -56
E. 64

35) Angle A and angle B are supplementary. The measure of angle A is 2 times the measure of angle B. What is the measure of angle A in degrees?

A. 100°
B. 120°
C. 140°
D. 160°
E. 170°

36) If $x=-2$ in the following equation, what is the value of y? $2x+3=\frac{y+6}{5}$

A. -9
B. -11
C. -13
D. -15
E. -17

37) Tomas is 6 feet 8.5 inches tall, and Alex is 5 feet 3 inches tall. What is the difference in height, in inches, between Alex and Tomas?

A. 2.5
B. 7.5
C. 12.5
D. 17.5
E. 19.5

38) Simplify: $\frac{\left(\frac{40(x+1)}{4}\right)-10}{12}$

A. $\frac{3}{7}x$
B. $\frac{5}{6}x$
C. $\frac{5}{12}x$
D. $\frac{12}{5}x$
E. $\frac{12}{7}x$

39) Yesterday Kylie writes 10% of her homework. Today she writes another 18% of the entire homework. What fraction of the homework is left for her to write?

A. $\frac{4}{25}$
B. $\frac{7}{25}$
C. $\frac{10}{25}$
D. $\frac{18}{25}$
E. $\frac{21}{25}$

40) In a box of blue and yellow pens, the ratio of yellow pens to blue pens is $2:3$. If the box contains 9 blue pens, how many yellow pens are there?

A. 2
B. 3
C. 4
D. 5
E. 6

41) What decimal is equivalent to $-\frac{6}{9}$?

A. $-0.\overline{5}$
B. $-0.\overline{6}$
C. $-0.\overline{65}$
D. $-0.\overline{7}$
E. $-0.\overline{75}$

42) The area of a circle is 81π. What is the diameter of the circle?

Write your answer in the box below.

43) Five years ago, Amy was three times as old as Mike was. If Mike is 10 years old now, how old is Amy?

A. 4
B. 8
C. 12
D. 14
E. 20

44) How many positive even factors of 68 are greater than 26 and less than 60?

A. 0
B. 1
C. 2
D. 4
E. 6

45) The ratio of two sides of a parallelogram is $2:3$. If its perimeter is $40\ cm$, find the length of its sides.

A. $6\ cm, 12\ cm$
B. $8\ cm, 12\ cm$
C. $10\ cm, 14\ cm$
D. $12\ cm, 16\ cm$
E. $14\ cm, 18\ cm$

46) What is the value of x in the following equation? $\frac{3}{4}(x-2) = 3(\frac{1}{6}x - \frac{3}{2})$

A. $\frac{1}{4}$
B. $-\frac{3}{4}$
C. -3
D. 6
E. -12

47) If x can be any integer, what is the greatest possible value of the expression $2 - x^2$?

A. -1
B. 0
C. 2
D. 3
E. 4

48) A store has a container of handballs: 6 green, 5 blue, 8 white, and 10 yellow. If one ball is

picked from the container at random, what is the probability that it will be green?

A. $\frac{1}{5}$
B. $\frac{6}{11}$
C. $\frac{6}{29}$
D. $\frac{8}{25}$
E. $\frac{11}{25}$

49) In the figure below, F is the midpoint of EH. Which segment has length $2y - x$ centimeters?

A. EF
B. GH
C. EG
D. FH
E. EG

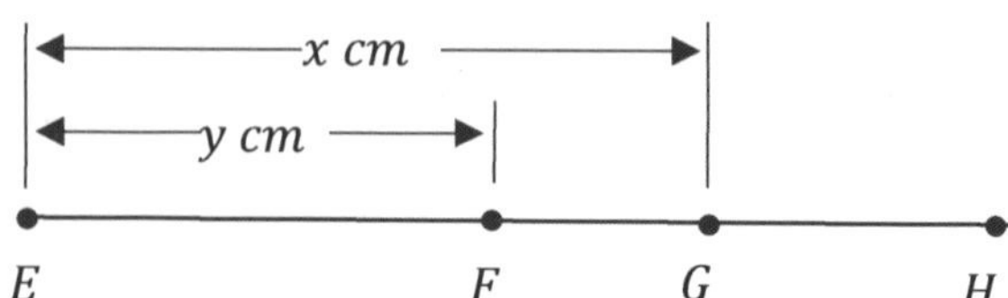

50) Emma answered 9 out of 45 questions on a test incorrectly. What percentage of the questions did she answer correctly?

A. 10%
B. 40%
C. 68%
D. 80%
E. 92%

51) If 30% of a number is 12, what is the number?

A. 12
B. 25
C. 40
D. 45
E. 50

52) If Anna multiplies her age by 5 and then adds 3, she will get a number equal to her mother's age. If x is her mother's age, what is Anna's age in terms of x?

A. $\frac{x-3}{5}$
B. $\frac{x-5}{3}$
C. $3x + 5$
D. $5x - 3$
E. $x - 3$

53) Jason is 15 miles ahead of Joe running at 4.5 miles per hour and Joe is running at the speed of 7 miles per hour. How long does it take Joe to catch Jason?

A. $3\ hours$
B. $4\ hours$
C. $6\ hours$
D. $8\ hours$
E. $10\ hours$

54) The reciprocal of $\frac{3}{5}$ is added to the reciprocal of $\frac{1}{4}$. What is the reciprocal of this sum?

A. $\frac{3}{5}$
B. $\frac{5}{3}$
C. $\frac{3}{17}$
D. $\frac{17}{3}$
E. $\frac{19}{3}$

55) In the following parallelogram, find the value of x?

Write your answer in the box below.

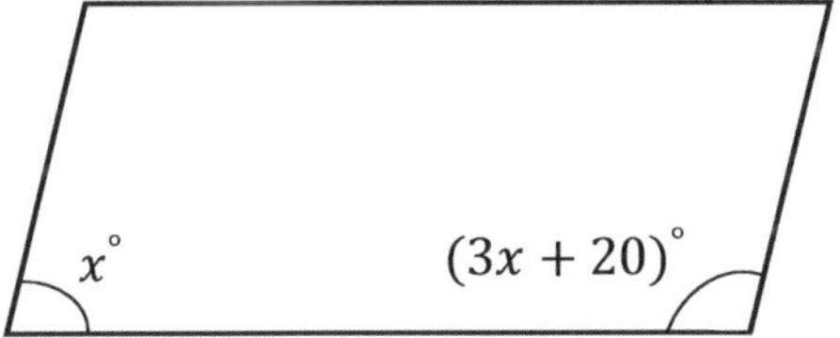

56) What is the solution to $\frac{0.02}{0.25} = \frac{1.25}{x}$?

A. 0.150
B. 1.156
C. 11.565
D. 15.625
E. 16.625

End of Praxis Mathematics Practice Test 2

Praxis Core Practice Tests Answer Keys

Now, it's time to review your results to see where you went wrong and what areas you need to improve.

Praxis Practice Test 1						Praxis Practice Test 2					
1	D	**21**	B	**41**	A	**1**	A	**21**	A	**41**	B
2	C	**22**	D	**42**	C	**2**	E	**22**	C	**42**	18
3	E	**23**	A	**43**	D	**3**	A	**23**	C	**43**	E
4	D	**24**	D	**44**	C	**4**	B	**24**	E	**44**	B
5	E	**25**	B	**45**	C	**5**	D	**25**	E	**45**	B
6	C	**26**	C	**46**	C	**6**	B	**26**	45	**46**	E
7	D	**27**	A	**47**	C	**7**	C	**27**	B	**47**	C
8	C	**28**	E	**48**	38	**8**	B	**28**	B	**48**	C
9	16	**29**	3.44	**49**	D	**9**	C	**29**	C	**49**	B
10	C	**30**	D	**50**	A	**10**	C	**30**	D	**50**	D
11	B	**31**	B	**51**	E	**11**	A	**31**	A	**51**	C
12	D	**32**	D	**52**	E	**12**	12	**32**	D	**52**	A
13	D	**33**	A	**53**	B	**13**	B	**33**	E	**53**	C
14	E	**34**	18	**54**	B	**14**	C	**34**	A	**54**	C
15	A	**35**	C	**55**	C	**15**	C	**35**	B	**55**	40
16	C	**36**	E	**56**	B	**16**	D	**36**	B	**56**	D
17	D	**37**	D			**17**	B	**37**	D		
18	D	**38**	E			**18**	C	**38**	B		
19	C	**39**	46			**19**	B	**39**	D		
20	B	**40**	B			**20**	275	**40**	E		

Praxis Core Practice Tests Answers and Explanations

Praxis Core Mathematics Practice Test 1 Answers and Explanations

1) Choice D is correct

The capacity of a red box is 20% bigger than the capacity of a blue box and it can hold 30 books. Therefore, we want to find a number that 20% bigger than that number is 30. Let x be that number. Then: $1.20 \times x = 30$. Divide both sides of the equation by 1.2. Then: $x = \frac{30}{1.20} = 25$

2) Choice C is correct

Convert everything into an equation: $35 = (3 \times \text{shirt}) - 10$

Now, solve the equation: $45 = 3 \text{ shirt} \rightarrow \text{shirt} = \frac{45}{3} = 15$. The price of the shirt was $\$15$.

3) Choice E is correct

First, convert the improper fraction to a mixed number: $-\frac{32}{5} = -6\frac{2}{5}$

The two closest integers to this fraction are -7 and -6.

The integer less than $-\frac{32}{5}$ is -7.

4) Choice D is correct

Let x equal the smallest angle of the triangle. Then, the three angles are $x, 3x$, and $5x$. The sum of the angles of a triangle is 180. Set up an equation using this to find x:

$x + 3x + 5x = 180 \rightarrow 9x = 180 \rightarrow x = 20$

Since the question asks for the measure of the largest angle, $5x = 5(20) = 100^\circ$

5) Choice E is correct

The angle $(2x - 5)$ and 55 are supplementary angles. Therefore:

$(2x - 5) + 55 = 180 \rightarrow 2x + 50 = 180 \rightarrow 2x = 180 - 50 \rightarrow 2x = 130 \rightarrow x = \frac{130}{2} \rightarrow x = 65$

6) Choice C is correct

There are 6 digits in the repeating decimal (0.142857), so digit 1 would be the first, seventh, thirteenth digit and so on. To find the 68th digit, divide 68 by 6.

$68 \div 6 = 11r2$

7) Choice D is correct

Solve for x. $-4 \leq 4x - 8 < 16 \Rightarrow$ Add 8 to all sides: $-4 + 8 \leq 4x - 8 + 8 < 16 + 8 \Rightarrow$

$4 \leq 4x < 24 \Rightarrow$ Divide all sides by 4: $1 \leq x < 6$. Choice D represents this inequality.

8) Choice C is correct

Based on triangle similarity theorem: $\frac{a}{a+b} = \frac{c}{3} \rightarrow c = \frac{3a}{a+b} = \frac{3\sqrt{3}}{3\sqrt{3}} = 1 \rightarrow$ Area of shaded region is: $\left(\frac{c+3}{2}\right)(b) = \frac{4}{2} \times 2\sqrt{3} = 4\sqrt{3}$

9) The answer is 16

Based on triangle similarity theorem, set up a proportion to solve for x:

$\frac{x+8}{x} = \frac{6}{4} \rightarrow 4(x+8) = 6x \rightarrow 4x + 32 = 6x \rightarrow 32 = 2x \rightarrow x = 16$

10) Choice C is correct

A linear equation is a relationship between two variables, x and y, and can be written in the form of $y = mx + b$. A non-proportional linear relationship takes on the form $y = mx + b$, where $b \neq 0$ and its graph is a line that does not cross through the origin. Only in graph C, the line does not pass through the origin.

11) Choice B is correct

Use simple interest formula:

$I = prt \ (I = interest\ , p = principal, r = rate, t = time)$

$I = prt \rightarrow 600 = (3{,}000)(0.05)(t) \rightarrow 600 = 150t \rightarrow t = 4$

12) Choice D is correct

To solve, add the two given fractions: $2\frac{2}{5} + 1\frac{3}{4}$

The common denominator is 20: $2\frac{8}{20} + 1\frac{15}{20} = 3\frac{23}{20} = 4\frac{3}{20}$

13) Choice D is correct

Consider the case where $k = 1$

$n - k = 46 \rightarrow n - 1 = 46 \rightarrow n - 1 + 1 = 46 + 1 \rightarrow n = 47$

The list of integers from 1 to 47 contains 47 numbers.

14) Choice E is correct

The original piece of paper is $2\frac{3}{5}$ feet long.

The shorter piece is x feet long, and it must be less than half the length of the original piece of paper. Since half of $2\frac{3}{5}$ is $1\frac{3}{10}$ it follows that $x < 1\frac{3}{10}$.

15) Choice A is correct

First, find the sum of course grade of Anna, $average = \frac{sum\ of\ terms}{number\ of\ terms} \Rightarrow$

$80 = \frac{sum\ of\ course\ grade}{5} \rightarrow the\ sum\ of\ course\ grade = 80 \times 5 = 400$

Anna and William have the same sum of course grade, now find the Williams mean

$$average = \frac{sum\ of\ course\ grade}{number\ of\ course} \Rightarrow \frac{400}{8} = 50$$

16) Choice C is correct

Since both ratios have y in common, solve for x and z in terms of y in both equations. Using $x: y = 1: 3$, solve for x in terms of y. $\frac{x}{y} = \frac{1}{3} \rightarrow x = \frac{1}{3}y$

Using the ratio $y: z = 2: 5$, solve for z in terms of y: $\frac{y}{z} = \frac{2}{5} \rightarrow z = \frac{5}{2}y$

The question states $x + y + z = 69$

Substitute from the two equations above and solve for y.

$\frac{1}{3}y + y + \frac{5}{2}y = 69 \rightarrow \frac{2y+6y+15y}{6} = 69 \rightarrow \frac{23}{6}y = 69 \rightarrow 23y = 414 \rightarrow y = 18$

17) Choice D is correct

Multiply each term by 3 to eliminate the fraction, and isolate x:

$-1(3) < \left(\frac{x}{3}\right)(3) < 2(3) \rightarrow -3 < x < 6$, therefore, x must be between -3 and 6. Only Choice D represents all values of x.

18) Choice D is correct

List in order the odd numbers between 5 to 30: 7,9,11,13,15,17,19,21,23,25,27, and 29. Since, the numbers are consecutive odd numbers, the mean and the median are equal. The median is the number in the middle. Since we have 12 numbers, the median is the average of numbers 6 and 7 which are 17 and 19. The mean (or the median) is: Mean $= \frac{17+19}{2} = 18$

19) Choice C is correct

Square both sides of the equation: $\left(\sqrt{2y}\right)^2 = \left(\sqrt{5x}\right)^2 \rightarrow 2y = 5x$

Solve for x: $x = \frac{2}{5}y$

20) Choice B is correct

The area of a trapezoid can be determined using the formula: $A = \frac{1}{2} \times (a + b) \times h$

We know: $DC = 4\ cm$, $AE = 2\ cm$, and $AD = 4\ cm \rightarrow$

$A = \frac{1}{2} \times (4\ cm + 2\ cm) \times 4\ cm = 12\ cm^2$

21) Choice B is correct

60% of students can't swim$\rightarrow 100 - 60 = 40\%$ can swim.

Then: $0.40 \times 45 = 18$

22) Choice D is correct

Money received by 14 person $= \$8{,}400$. So, the money received by one person is:

$\frac{\$8,400}{14} = \$600$

23) Choice A is correct

There are 12 sticks in the box $(6 + 4 + 2)$. So, the probability that Emma picks a green stick is: $Probability = \frac{6}{12} = \frac{1}{2}$

24) Choice D is correct

First, calculate the area of the circle and the area of the square:

Area of the circle $= \pi r^2 = \pi(6)^2 = 36\pi = 113.04\ cm^2$

Area of the square $= 5\ cm \times 5\ cm = 25\ cm^2$

To calculate the shaded area, subtract the area of the square from the area of the circle: $113.04\ cm^2 - 25\ cm^2 = 88.04\ cm^2$

25) Choice B is correct

Use the percent increase expression to find the answer:

$\frac{new\ price - original\ price}{original\ price} = \frac{5.67 - 5.40}{5.40} = 0.05 = 5\%$

26) Choice C is correct

Let x be the number of blue marbles. Write the items in the ratio as a fraction:

$\frac{x}{150} = \frac{4}{3} \rightarrow 3x = 600 \rightarrow x = 200$

27) Choice A is correct

Let x be the number: $\frac{5}{8}x = 90 \rightarrow x = 90 \times \frac{8}{5} = \frac{720}{5} = 144$

28) Choice E is correct

Set up a proportion to solve: $\frac{\frac{5}{14}\ cherry}{\frac{5}{70}\ apple} = \frac{x\ cherry}{1\ apple} \rightarrow \frac{5}{14} \times \frac{70}{5} = x \rightarrow x = \frac{70}{14} = \frac{10}{2} \rightarrow x = 5$

29) The answer is 3.44

The area of square $LMNO$ is 16 square centimeters.

So: $S^2 = 16 \rightarrow \sqrt{S^2} = \sqrt{16} \rightarrow S = 4\ cm$

Sides LO and NO are each a radius of the circle. So, the radius of the circle is $4\ cm$.

calculate the area of $\frac{1}{4}$ of the circle. The area of a circle is $A = \pi r^2$. So the area of the $\frac{1}{4}$ of the circle, in square centimeters, is $\frac{1}{4}A = \frac{1}{4}\pi r^2 = \frac{1}{4}\pi(4)^2 = \frac{1}{4}\pi(16) = 4\pi$. For calculating the area of shaded region, subtract area of $\frac{1}{4}$ of the circle from area of square. The area of the shaded par: $16 - 4\pi$: and $\pi = 3.14$, then, the answer is:

$16 - 4\pi = 16 - 12.56 = 3.44$

30) Choice D is correct

$2m = n + 5 \rightarrow m = \frac{n+5}{2}$. Substitute each value of n to find the values of m:

$$m = \frac{5+5}{2} = \frac{10}{2} = 5$$

$$m = \frac{3+5}{2} = \frac{8}{2} = 4$$

$$m = \frac{7+5}{2} = \frac{12}{2} = 6$$

The set of m is $\{5,4,6\}$.

31) Choice B is correct

Plug in 25 for x in the equation.

A. $x + 10 = 40 \rightarrow 25 + 10 \neq 40$

B. $4x = 100 \rightarrow 4(25) = 100$

C. $3x = 70 \rightarrow 3(25) \neq 70$

D. $\frac{x}{2} = 12 \rightarrow \frac{25}{2} \neq 12$

E. $\frac{x}{3} = 8 \rightarrow \frac{25}{3} \neq 8$

Only choice B is correct.

32) Choice D is correct

Jack scored a mean of 80 per test. In the first 4 tests, the sum of scores is:

$80 \times 4 = 320$. Now, calculate the mean over the 5 tests: $\frac{320+90}{5} = \frac{410}{5} = 82$

33) Choice A is correct

Volume of the cube is less than $64\ m^3$. Use the formula of volume of cubes.

$Volume = (one\ side)^3 \Rightarrow 64 = (one\ side)^3$. Find the cube root of both sides.

$64 = (one\ side)^3 \rightarrow one\ side = \sqrt[3]{64} = 4\ m$

Then: $4 =$ one side. The side of the cube is less than 4. Only choice A is less than 4.

34) The answer is 18

First draw an isosceles triangle. Remember that two legs of the triangle are equal.

Let put a for the legs. Then:

$a = 6 \Rightarrow$Area of the triangle is $= \frac{1}{2}(6 \times 6) = \frac{36}{2} = 18\ cm^2$

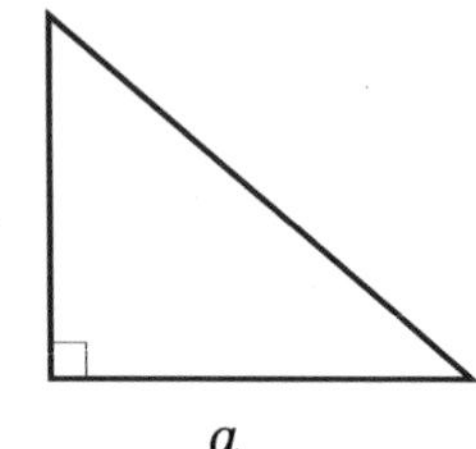

35) Choice C is correct

Solve for x: $0.00104 = \frac{104}{x}$, multiply both sides by x, $(0.00104)(x) = \frac{104}{x}(x)$.

Simplify: $0.00104x = 104$. Divide both side by 0.00104: $\frac{0.00104x}{0.00104} = \frac{104}{0.00104}$, simplify

$x = \frac{104}{0.00104} = 100{,}000$

36) Choice E is correct

The number of cards in the bag is 15.

$Probability = \frac{number\ of\ desired\ outcomes}{number\ of\ total\ outcomes} = \frac{1}{15}$

37) Choice D is correct

The equation of a line is in the form of $y = mx + b$, where m is the slope of the line and b is the $y - intercept$ of the line. Two points $(4, 3)$ and $(3, 2)$ are on line A. Therefore, the slope of the line A is: $slope\ of\ line\ A = \frac{y_2 - y_1}{x_2 - x_1} = \frac{2-3}{3-4} = \frac{-1}{-1} = 1$

The slope of line A is 1. Thus, the formula of the line A is: $y = mx + b = x + b$, choose a point and plug in the values of x and y in the equation to solve for b. Let's choose point $(4, 3)$. Then: $y = x + b \rightarrow 3 = 4 + b \rightarrow b = 3 - 4 = -1$

The equation of line A is: $y = x - 1$

Now, let's review the choices provided:

A. $(5, 7)$ $y = x - 1 \rightarrow 7 = 5 - 1 \rightarrow 7 \neq 4$ This is not true.

B. $(3, 4)$ $y = x - 1 \rightarrow 4 = 3 - 1 \rightarrow 4 \neq 2$ This is not true.

C. $(-1, 2)$ $y = x - 1 \rightarrow 2 = -1 - 1 \rightarrow 2 \neq -2$ This is not true.

D. $(-1, -2)$ $y = x - 1 \rightarrow -2 = -1 - 1 \rightarrow -2 = -2$ This is true.

E. $(-7, -9)$ $y = x - 1 \rightarrow -9 = -7 - 1 \rightarrow -9 \neq -8$ This is not true.

38) Choice E is correct

$g(x) = -2$, then $f(g(x)) = f(-2) = 2(-2)^3 + 5(-2)^2 + 2(-2) = -16 + 20 - 4 = 0$

39) The answer is 46

$(2x + 4)^{\circ}$ and 96° are vertical angles. Vertical angles are equal in measure.

Then: $2x + 4 = 96 \rightarrow 2x = 92 \rightarrow x = 46$

40) Choice B is correct

The two-digit numbers must be even, so the only possible two-digit numbers must end in 6, since 6 is the only even digit given in the problem. Since the numbers cannot be repeated, the only possibilities for two-digit even numbers are 76 and 56. Thus, the answer is two possible two-digit numbers.

41) Choice A is correct

First convert 24 inches to feet. 12 inch $= 1$ feet, thus: $24 \div 12 = 2\ feet$. Then, calculate the volume, in cubic feet: $25 \times 6 \times 2 = 300\ ft^3$

42) Choice C is correct

Use properties of equations to determine the missing expression. $\frac{2y}{x} - \frac{y}{3x} = \frac{(\dots)}{3x}$

$\frac{3}{3} \cdot \frac{2y}{x} - \frac{y}{3x} = \frac{(\dots)}{3x} \rightarrow \frac{6y}{3x} - \frac{y}{3x} = \frac{(\dots)}{3x} \rightarrow \frac{6y-y}{3x} = \frac{(\dots)}{3x} \rightarrow (\dots) = 5y$

43) Choice D is correct

First calculate exponents value, then multiplying and subtracting:

$200(3 + 0.01)^2 - 200 = 200(3.01)^2 - 200 = 200(9.0601) - 200 = 1{,}612.02$

44) Choice C is correct

Since 90 boxes contain $360\ kg$ vegetable. Therefore, 1 box contains $\frac{360\ kg}{90} = 4\ kg$ vegetable.

45) Choice C is correct

Let n represent a number in the sequence, and let x represent the number that comes just before $n. n = 4 + 2x \rightarrow 84 = 4 + 2x \rightarrow 80 = 2x \rightarrow x = 40$

46) Choice C is correct

Use PEMDAS (order of operation): $[6 \times (-24) + 8] - (-4) + [4 \times 5] \div 2 =$

$[-144 + 8] - (-4) + [20] \div 2 = [-144 + 8] + 4 + 10 = [-136] + 4 + 10 = -122$

47) Choice C is correct

First, find a common denominator for 2 and $\frac{3x}{x-5}$. It's $x - 5$. Then:

$2 + \frac{3x}{x-5} = \frac{2(x-5)}{x-5} + \frac{3x}{x-5} = \frac{2x-10+3x}{x-5} = \frac{5x-10}{x-5}$. Now, multiply the numerator and denominator of $\frac{3}{5-x}$ by -1. Then: $\frac{3\times(-1)}{(5-x)\times(-1)} = \frac{-3}{x-5}$. Rewrite the expression: $\frac{5x-10}{x-5} = \frac{-3}{x-5}$. Since the denominators of both fractions are equal, then, the numerators must be equal.

$5x - 10 = -3 \rightarrow 5x = 7 \rightarrow x = \frac{7}{5}$.

48) The answer is 38

Perimeter of rectangle is equal to the sum of all the sides of the rectangle:

Perimeter $= 2(14) + 2(5) = 28 + 10 = 38\ cm$

49) Choice D is correct

$3\frac{1}{4} + 2\frac{4}{16} + 1\frac{3}{8} + 5\frac{1}{2}$. Convert all the fractions to a common denominator (16):

$3\frac{4}{16} + 2\frac{4}{16} + 1\frac{6}{16} + 5\frac{8}{16} = (3 + 2 + 1 + 5) + \left(\frac{4+4+6+8}{16}\right) = 11 + 1\frac{6}{16} = 12\frac{6}{16} = 12\frac{3}{8}$

50) Choice A is correct

One gram is equal to 1,000 milligrams, or 1 milligram is equal to $\frac{1}{1,000}$ gram.

Thus, 85 milligrams $= \frac{85}{1,000} = 0.085$ gram

51) Choice E is correct

Check each choice provided:

A. 1 $\quad \frac{4+5+8+11+12}{5} = \frac{40}{5} = 8$

B. 4 $\quad \frac{1+5+8+11+12}{5} = \frac{37}{5} = 7.4$

C. 5 $\quad \frac{1+4+8+11+12}{5} = \frac{36}{5} = 7.2$

D. 8 $\quad \frac{1+4+5+11+12}{5} = \frac{33}{5} = 6.6$

E. 11 $\quad \frac{1+4+5+8+12}{5} = \frac{30}{5} = 6$

52) Choice E is correct

Substitute 6 for m and -3 for n:

$\frac{5-9(3+n)}{3m-5(2-n)} = \frac{5-9(3+(-3))}{3(6)-5(2-(-3))} = \frac{5-9(0)}{18-5(5)} = \frac{5}{18-25} = \frac{5}{-7} = -\frac{5}{7}$

53) Choice B is correct

To answer this question, we need to divide 28 by 7: $\frac{28}{7} = 4$

54) Choice B is correct

Simple interest (y) is calculated by multiplying the initial deposit (p), by the interest rate (r), and time (t). $360 = 240 \times 0.10 \times t \rightarrow 360 = 24t \rightarrow t = \frac{360}{24} = 15$

So, it takes 15 years to get $360 with an investment of $240.

55) Choice C is correct

There are 60 minutes in 1 hours. Divide the number of minutes by the number of minutes in 1 hour: $\frac{1,800}{60} = 30$ hours

56) Choice B is correct

First calculate absolute values: $|-44| - x + |8| = 36 \rightarrow 44 - x + 8 = 36$

Combine like terms: $44 + 8 - 36 = x \rightarrow x = 16$

Praxis Core Mathematics Practice Test 2 Answers and Explanations

1) Choice A is correct

First, find the sale price. 15% of \$45.00 is \$6.75, so the sale price is

$\$45.00 - \$6.75 = \$38.25$. Next, find the price after Nick's employee discount. $20\% \times \$38.25 = \$7.65$, so, the final price of the shoes is $\$38.25 - \$7.65 = \$30.60$.

2) Choice E is correct

Write the numbers in the ratio and simplify: $18:12 \rightarrow 3:2$

3) Choice A is correct

The area of the entire circle is πr^2. The fraction of the circle that is shaded is $\frac{30}{360} = \frac{1}{12}$. So, the area of the sector is $\frac{1}{12}\pi r^2$. Use that information to find r.

$$\frac{1}{12}\pi r^2 = 12\pi \rightarrow r^2 = 144 \rightarrow r = 12$$

Use r to calculate the circumference of the circle: $c = 2\pi r = 2\pi(12) = 24\pi$

The circumference is $24\pi\ feet$.

4) Choice B is correct

The factors of 45 are: $\{1, 3, 5, 9, 15, 45\}$. Only choice B is correct.

5) Choice D is correct

$$Percent\ of\ change = \frac{new\ number - original\ number}{original\ number} = \frac{18.36 - 15.30}{15.30} = 20\%$$

6) Choice B is correct

First find the multiples of 8 that fall between 10 and 60: $16, 24, 32, 40, 48, 56$. Since the greatest common factor of 32 and x is 8, x cannot be 32 (otherwise the GCF would be 32, not 8). There are 5 remaining values: $16, 24, 40, 48$ and 56. Number 16 is also not possible (otherwise the GCF would be 16, not 8). Then, there are 4 possible values for x.

7) Choice C is correct

The total number of candies in the box is $6 + 4 + 3 = 13$. The number of candies that are not orange is $6 + 3 = 9$. The probability of the first candy not being orange is $\frac{9}{13}$. Now, out of 12 candies, there are 8 candies left that are not orange. The probability of the second candy not being orange is $\frac{8}{12}$. Multiply these two probabilities to get the solution: $\frac{9}{13} \times \frac{8}{12} = \frac{72}{156} = \frac{24}{52} = \frac{6}{13}$

8) Choice B is correct

First, change the improper fractions into mixed numbers: $\frac{7}{2} = 3\frac{1}{2}$ and $\frac{30}{4} = 7\frac{1}{2}$

The integers between these two values are $4, 5, 6$ and 7. So, there are 4 integers between $\frac{7}{5}$ and $\frac{30}{4}$.

9) Choice C is correct

The increase in sales tax percentage is $8.5\% - 8.0\% = 0.5\%$

0.5% of $\$250$ is $(0.5\%)(250) = (0.005)(250) = 1.25\$$

10) Choice C is correct

When a point is reflected over x axes, the (y) coordinate of that point changes to $(-y)$ while its x coordinate remains the same. $C(7, 9) \rightarrow C'(7, -9)$

11) Choice A is correct

$2x + 4 > 11x - 12.5 - 3.5x \rightarrow$ Combine like terms: $2x + 4 > 7.5x - 12.5 \rightarrow$ Subtract $2x$ from both sides: $4 > 5.5x - 12.5$. Add 12.5 both sides of the inequality. $16.5 > 5.5x$, Divide both sides by $5.5 \rightarrow \frac{16.5}{5.5} > x \rightarrow x < 3$

12) The answer is 12

Use the volume of the triangular prism formula.

$$V = \frac{1}{2}(length)(base)(high) \rightarrow V = \frac{1}{2} \times 4 \times 3 \times 2 \Rightarrow V = 12\ m^3$$

13) Choice B is correct

First, list the products:

$1 \times 4 = 4$

$1 \times 6 = 6$

$1 \times 5 = 5$

$1 \times 7 = 7$

$4 \times 6 = 24$

$4 \times 5 = 20$

$4 \times 7 = 28$

$6 \times 5 = 30$

$6 \times 7 = 42$

$5 \times 7 = 35$

Out of 10 results, 3 numbers are odd. The answer is: $\frac{3}{10}$

14) Choice C is correct

Standard form of straight-line equation is: $y = mx + b$. Thus, choice C has a graph that is a straight line. All other options are not equations of straight lines.

15) Choice C is correct

Since $5n$ is even, then $5n + 1$ must be odd. Thus $5n + 3$ and $5n + 5$ are also odd. So, there are a total of 3 numbers in this range that are odd.

16) Choice D is correct

Isolate and solve for x: $\frac{2}{3}x + \frac{1}{6} = \frac{1}{3} \Rightarrow \frac{2}{3}x = \frac{1}{3} - \frac{1}{6} \Rightarrow \frac{2}{3}x = \frac{1}{6}$

Multiply both sides by the reciprocal of the coefficient of x.

$\left(\frac{3}{2}\right)\frac{2}{3}x = \frac{1}{6}\left(\frac{3}{2}\right) \Rightarrow x = \frac{3}{12} = \frac{1}{4}$

17) Choice B is correct

Use simple interest formula: $I = prt$,

$(I = interest, p = principal, r = rate, t = time) \rightarrow I = (16{,}000)(0.025)(3) = 1{,}200$

18) Choice C is correct

First, use the given information to calculate the value of a: $\frac{a}{4} = b \rightarrow \frac{a}{4} = 2 \rightarrow a = 8$

Now, calculate $a^2 + 4b$ by substituting $a = 8$ and $b = 2$, $\quad (8)^2 + 4(2) = 72$

19) Choice B is correct

Write the ratios in fraction form and solve for x: $\frac{x}{40} = \frac{20}{32}$

Cross multiply: $32x = 800$. Apply the multiplicative inverse property; divide both sides by 32: $x = \frac{800}{32} = 25$

20) The answer is 275

76% of the people polled answered yes, so 24% of the people did not answer yes. Therefore, 66 people is 24% of the total, x.

$\frac{66}{x} = \frac{24}{100} \rightarrow \frac{66}{x} = \frac{6}{25} \rightarrow 66(25) = 6x \rightarrow \frac{66(25)}{6} = x \rightarrow x = 275$

21) Choice A is correct

Since 0.0099 is equal to 0.99%, the closest to that value is 0.1%.

22) Choice C is correct

Surface Area of a cylinder $= 2\pi r(r + h)$,The radius of the cylinder is 3 $(6 \div 2)$ inches and its height is 8 inches.

Therefore, Surface Area of a cylinder $= 2\pi(3)(3+8) = 66\pi\ in^2$

23) Choice C is correct

First, find the speed of the train in miles per hour: $280 \div 4 = 70$ miles per hour

The number of miles left to travel is: $1{,}500 - 280 = 1{,}220$ miles

To find the number of hours left, use the equation

$d = rt \rightarrow (distance) = (rate) \times (time) \rightarrow 1{,}220 = 70t$

$t = \frac{1{,}220}{70} = 17.4285714$ hours. That number rounded to the nearest whole hour is 17 hours.

24) Choice E is correct

Apply absolute equation rule $-3 \leq 9 + 6x \leq 3$. Add -9 to all sides. Then:

$-3 - 9 \leq 9 + 6x - 9 \leq 3 - 9 \ \rightarrow -12 \leq 6x \leq -6$. Now, divide all sides by 6:

$-2 \leq x \leq -1$. Choice E represents this inequality.

25) Choice E is correct

The line passes through the origin, $(6, m)$ and $(m, 12)$. Any two of these points can be used to find the slope of the line. Since the line passes through $(0, 0)$ and $(6, m)$, the slope of the line is equal to $\frac{m-0}{6-0} = \frac{m}{6}$. Similarly, since the line passes through $(0, 0)$ and $(m, 12)$, the slope of the line is equal to $\frac{12-0}{m-0} = \frac{12}{m}$. Since each expression gives the slope of the same line, it must be true that $\frac{m}{6} = \frac{12}{m}$,Using cross multiplication gives

$$\frac{m}{6} = \frac{12}{m} \rightarrow m^2 = 72 \rightarrow m = \pm\sqrt{72} = \pm\sqrt{36 \times 2} = \pm\sqrt{36} \times \sqrt{2} = \pm 6\sqrt{2}$$

26) The answer is 45

Perimeter of figure A is: $2\pi r = 2\pi \frac{10}{2} = 10\pi = 10 \times 3 = 30$

Area of figure B is: $6 \times 10 = 60$, Average $= \frac{30+60}{2} = \frac{90}{2} = 45$

27) Choice B is correct

Let x be the original price. Then:

$\$1{,}912.50 = x - 0.15(x) \rightarrow 1{,}912.50 = 0.85x \rightarrow x = \frac{1{,}912.50}{0.85} \rightarrow x = 2{,}250$

28) Choice B is correct

Given diameter $= 18\ cm \rightarrow$ radius $= 9\ cm$

Perimeter of circle $= 2\pi r = 2 \times \pi \times 9 = 18\pi = 56.52\ cm$

The perimeter of semi−circular slice is:

$P = \frac{perimeter\ of\ circle}{2} + 2r = \frac{56.52}{2} + 2(9) = 46.26\ cm$

29) Choice C is correct

Let x be the number of inches representing 2.5 feet. Set up a proportion and solve for x: $\frac{x}{2.5} = \frac{0.15}{150} \rightarrow x = \frac{0.15\times2.5}{150} \rightarrow x = 0.0025\ in$

30) Choice D is correct

Set an equation: $\frac{3}{7}Z = 54$

Solve for Z: $\rightarrow Z = 54 \times \frac{7}{3} = 126,$ then, calculate $\frac{2}{5}Z$: $\frac{2}{5} \times 126 = 50.4$

31) Choice A is correct

To answer this question, multiply 72 miles per hour to $8 \rightarrow 72 \times 8 = 576$ miles

32) Choice D is correct

The spent amount is \$60, and the tip is 25%. Then: $tip = 0.25 \times 60 = \$15$

Final price = Selling price+tip → final price $= \$60 + \$15 = \$75$

33) Choice E is correct

To find the smallest factor of 95, list the factors: $1, 5, 19,$ and 95.The smallest factor (other than 1) is 5. Of the choices listed $(28, 32, 39,$ and $45)$, only 45 is a multiple of 5.

34) Choice A is correct

Adding exponents is done by calculating each exponent first and then adding and dividing:

$\frac{4^2+3^2+(-5)^2}{(9+10-11)^2} = \frac{16+9+25}{(8)^2} = \frac{50}{64} = \frac{25}{32}$

35) Choice B is correct

Angle A and angle B are supplementary, so the sum of their angles is 180°.

Let a equal the measure of angle A, and let b equal the measure of angle B.

$a + b = 180$

The measure of angle A is 2 times the measure of angle B.

$$a = 2b \rightarrow 2b + b = 180 \rightarrow 3b = 180 \rightarrow b = \frac{180}{3} = 60$$

$a = 2b = 2(60) = 120$

Therefore, the measure of angle A is 120°.

36) Choice B is correct

Substitute -2 for x in the equation: $2(-2) + 3 = \frac{y+6}{5} \rightarrow -1 = \frac{y+6}{5} \rightarrow$

$y + 6 = -5 \rightarrow y = -5 - 6 = -11$

37) Choice D is correct

First, convert their heights from feet and inches to inches, by multiplying the number of feet by 12 and adding the inches. Tomas:

6 feet +8.5 inches. 6(12 inches) +8.5 inches= 72 inches+8.5 inches= 80.5 inches.

Alex: 5 feet +3 inches. 5(12 inches)+3 inches $= 60$ inches $+ 3$ inches $= 63$ inches

Then, subtract Alex's height from Tomas's height: $80.5 - 63 = 17.5$

38) Choice B is correct

Divide 40 by 4: $\frac{\left(\frac{40(x+1)}{4}\right)-10}{12} \rightarrow \frac{10(x+1)-10}{12}$.

Distribute 10 through the parentheses $(x+1)$

$$\frac{10x+10-10}{12} = \frac{10x}{12} = \frac{5}{6}x$$

39) Choice D is correct

So far, Kylie has written $10\% + 18\% = 28\%$ of the entire homework. That means she has $100\% - 28\% = 72\%$ left to write. $72\% = \frac{72}{100} = \frac{18}{25}$

40) Choice E is correct

Let x be the number of yellow pens. Write a proportion and solve: $\frac{yellow}{blue} = \frac{2}{3} = \frac{x}{9}$

Solve the equation: $18 = 3x \rightarrow x = 6$

41) Choice B is correct

To find the decimal equivalent to $-\frac{6}{9}$, divide 6 by 9. Then: $-\frac{6}{9} = -0.66666 \ldots = -0.\overline{6}$

42) The answer is 18

The formula for the area of the circle is πr^2 , The area is 81π. Therefore:

$A = \pi r^2 \Rightarrow 81\pi = \pi r^2$, Divide both sides by π: $81 = r^2 \Rightarrow r = 9$, Diameter of a circle is $2 \times radius$. Then: $Diameter = 2 \times 9 = 18$

43) Choice E is correct

Five years ago, Amy was three times as old as Mike. Mike is 10 years now. Therefore, 5 years ago Mike was 5 years. Five years ago, Amy was: $A = 3 \times 5 = 15$, Now Amy is 20 years old: $15 + 5 = 20$

44) Choice B is correct

List the factors of 68: 1 and 68, 2 and 34, 4 and 17. There is one factor greater than 26 and less than 60.

45) Choice B is correct

Let the lengths of two sides of the parallelogram be $2x\ cm$ and $3x\ cm$ respectively. Then, its perimeter $= 2(2x + 3x) = 10x$

Therefore, $10x = 40 \rightarrow x = 4$

One side $= 2(4) = 8\ cm$ and other side is: $3(4) = 12\ cm$

46) Choice E is correct

Isolate x in the equation and solve. Then:

$\frac{3}{4}(x-2) = 3\left(\frac{1}{6}x - \frac{3}{2}\right)$, expand $\frac{3}{4}$ and 3 to the parentheses $\rightarrow \frac{3}{4}x - \frac{3}{2} = \frac{1}{2}x - \frac{9}{2}$. Add $\frac{3}{2}$ to both sides: $\frac{3}{4}x - \frac{3}{2} + \frac{3}{2} = \frac{1}{2}x - \frac{9}{2} + \frac{3}{2}$. Simplify: $\frac{3}{4}x = \frac{1}{2}x - 3$. Now, subtract $\frac{1}{2}x$ from both sides: $\frac{3}{4}x - \frac{1}{2}x = \frac{1}{2}x - 3 - \frac{1}{2}x$. Simplify: $\frac{1}{4}x = -3$. Multiply both sides by 4:

$(4)\frac{1}{4}x = -3(4)$, simplify $x = -12$

47) Choice C is correct

To answer this question, assign several positive and negative values to x and determine what the value of the expression will be:

x	-1	0	2	3	4
$2-x^2$	1	2	-2	-7	-14

So, the maximum value of the expression is 2.

48) Choice C is correct

The total number of handballs in the container is $6 + 5 + 8 + 10 = 29$. Since there are 6 green handballs, the probability of selecting a green handball is $\frac{6}{29}$.

49) Choice B is correct

These facts are given: F is the midpoint of EH.

EG has a length of x cm. EF has a length of y cm.

Use the first two facts to determine that E has a length of $2y$ centimeters:

$GH = EH - EG = 2y - x$

50) Choice D is correct

If Emma answered 9 out of 45 questions incorrectly, then she answered 36 questions correctly.
$\frac{36}{45} \times 100 = 80\%$

51) Choice C is correct

Let x be the number. Write the equation and solve for x.

30% of $x = 12 \Rightarrow 0.30x = 12 \Rightarrow x = 12 \div 0.30 = 40$

52) Choice A is correct

Let y be Anna's age: $5y + 3 = x \rightarrow 5y = x - 3 \rightarrow y = \frac{x-3}{5}$

53) Choice C is correct

The distance between Jason and Joe is 15 miles. Jason running at 4.5 miles per hour and Joe is running at the speed of 7 miles per hour. Therefore, every hour the distance is 2.5 miles less. $15 \div 2.5 = 6\ hours$

54) Choice C is correct

The reciprocal of $\frac{3}{5}$ is added to the reciprocal of $\frac{1}{4}$:

$\frac{4}{1} + \frac{5}{3} = \frac{12}{3} + \frac{5}{3} = \frac{17}{3}$. The reciprocal of this sum is $\frac{3}{17}$.

55) The answer is 40

In a parallelogram, consecutive angles are supplementary. Thus,

$x + (3x + 20) = 180 \rightarrow 4x + 20 = 180 \rightarrow 4x = 160 \rightarrow x = 40$

56) Choice D is correct

To eliminate the decimals in this equation, multiply the numerators and denominators by 100:

$\left(\frac{0.02}{0.25}\right)\left(\frac{100}{100}\right) = \left(\frac{1.25}{x}\right)\left(\frac{100}{100}\right) \rightarrow \left(\frac{2}{25}\right) = \frac{125}{100x} \rightarrow x = \left(\frac{125}{100}\right)\left(\frac{25}{2}\right) = 15.625$

Effortless Math's Praxis Core Online Center

... So Much More Online!

Effortless Math Online Praxis Core Math Center offers a complete study program, including the following:

- ✓ Step-by-step instructions on how to prepare for the Praxis Core Math test
- ✓ Numerous Praxis Core Math worksheets to help you measure your math skills
- ✓ Complete list of Praxis Core Math formulas
- ✓ Video lessons for Praxis Core Math topics
- ✓ Full-length Praxis Core Math practice tests
- ✓ And much more...

No Registration Required.

Visit **EffortlessMath.com/Praxis** to find your online Praxis Core Math resources.

Receive the PDF version of this book or get another FREE book!

Thank you for using our Book!

Do you LOVE this book?

Then, you can get the PDF version of this book or another book absolutely FREE!

Please email us at:

info@EffortlessMath.com

for details.

Author's Final Note

I hope you enjoyed reading this book. You've made it through the book! Great job!

First of all, thank you for purchasing this study guide. I know you could have picked any number of books to help you prepare for your Praxis Core Math test, but you picked this book and for that I am extremely grateful.

It took me years to write this study guide for the Praxis Core Math because I wanted to prepare a comprehensive Praxis Core Math study guide to help test takers make the most effective use of their valuable time while preparing for the test.

After teaching and tutoring math courses for over a decade, I've gathered my personal notes and lessons to develop this study guide. It is my greatest hope that the lessons in this book could help you prepare for your test successfully.

If you have any questions, please contact me at reza@effortlessmath.com and I will be glad to assist. Your feedback will help me to greatly improve the quality of my books in the future and make this book even better. Furthermore, I expect that I have made a few minor errors somewhere in this study guide. If you think this to be the case, please let me know so I can fix the issue as soon as possible.

If you enjoyed this book and found some benefit in reading this, I'd like to hear from you and hope that you could take a quick minute to post a review on the book's Amazon page. To leave your valuable feedback, please visit: amzn.to/3cOZvVE

Or scan this QR code.

I personally go over every single review, to make sure my books really are reaching out and helping students and test takers. Please help me help Praxis Core Math test takers, by leaving a review!

I wish you all the best in your future success!

Reza Nazari

Math teacher and author

Printed in Great Britain
by Amazon